UMTS Network Planning and Development

Design and Implementation of the 3G CDMA Infrastructure

Chris Braithwaite and Mike Scott

AMSTERDAM • BOSTON • HEIDELBERG • LONDON • NEW YORK • OXFORD
PARIS • SAN DIEGO • SAN FRANCISCO • SINGAPORE • SYDNEY • TOKYO
Newnes is an imprint of Elsevier

Newnes
An imprint of Elsevier
Linacre House, Jordan Hill, Oxford OX2 8DP
200 Wheeler Road, Burlington, MA 01803

First published 2004

British Library Cataloguing in Publication Data
Braithwaite, Chris
UMTS network planning and development : design and implementation of the 3G CDMA infrastructure
1. Universal Mobile Telecommunications System
I. Title II. Scott, Mike
621.3'8456

Library of Congress Cataloging-in-Publication Data
Braithwaite, Chris.
UMTS network planning and development : design and implementation of the 3G CDMA infrastructure / Chris Braithwaite and Mike Scott.
p. cm.
Includes bibliographical references and index.
ISBN 0-7506-6082-1 (alk. paper)
1. Universal Mobile Telecommunications System. 2. Code division multiple access. 3. Network computers. I. Scott, Mike. II. Title.

TK5103.4883.B73 2003
621.382'1–dc22 2003066754

ISBN 0 7506 6082 1

For information on all Elsevier Newnes publications visit our website at www.newnespress.com

Typeset by Charon Tec Pvt. Ltd, Chennai, India
Printed and bound in Meppel, The Netherlands by Krips bv.

Contents

List of Figures

Part III Figures

Part IV Figures

Part V Figures

Part VI Figures

List of Tables

Part I Tables

Part II Tables

Part III Tables

Part IV Tables

Part V Tables

Part VI Tables

Author Profiles

Chris Braithwaite HNC

Chris has been in the electronics and telecommunications industry for over 20 years. More recently Chris has designed, developed, and presented technical cellular telecom training programmes and courseware for various operators and vendors Europe-wide. He now also specializes in universal mobile telephony system (UMTS), particularly UMTS radio network planning and design.

Chris's technical expertise includes successfully implementing networks and services throughout Europe and the Middle East including Switzerland, Austria, Holland, and the Czech Republic. He has extensive experience in radio frequency (RF) network planning, technical implementation, commissioning and integration of all network elements within the cellular communications industry, and has worked for both vendors and operators throughout Europe. As a major contributor of such major contracts he has also provided training and monitoring as part of the handover to network operations staff. He possesses a wide range of technical expertise within the telecoms field (specifically DCS 1800 networks, including general packet radio service, GPRS, UMTS/UTRAN, wideband-code division multiple access, W-CDMA), and the design and development towards both 2.5G and 3G systems.

Mike Scott MSc, MBA MinstBA

Mike, who holds an MSc in Engineering Geomaterials, an MBA and membership of the Institute of Business Managers, has been involved in the development and implementation of telecom networks with regard to site planning and construction for more than 8 years. This network implementation has been aided by Mike's extensive construction, planning, quality, and commercial experience supported by both his vocational knowledge and academic studies. He has worked with O_2, Ericsson, Nokia, Orange, T-Mobile, Sony, Taylor Woodrow, British Telecom and Oskar (TIW) etc. Mike has also been involved with the writing of network implementation design guides for Ericsson, Orange and British Telecom.

Apart from the planning and implementation of the base stations Mike has also specialized in the continued development of 'Stealth Sites' in both National parks and Historical centres, along with the cost benefit analysis of strategic base station deployment.

Having worked as a consultant over the last 6 years in numerous countries within Europe including Italy, Hungary, Belgium, Slovakia and the Czech Republic, he has gained in-depth experience of the difficulties of planning networks and the realistic limitations that can occur both in the technical and the construction planning process. Furthermore, he has been involved in research, analysing the efficient balance of network planning with realistic construction alternatives both in term of design and acquisition.

Acknowledgements

It is not possible to thank everyone who has had an important influence on the conception along with the drive and encouragement devising the structure and contents of this book. Eternal thanks go to our partners Katerina and especially Erin who donated many hours in the final stages and compilation. Without their unselfish love and support together with the patience of our children Natasha, Max and Dominika, this book would not have been possible. Therefore this book has been dedicated to them. Many thanks also go to our good friends Ray Gibson and Dr Anne Scott, who both provided endless support and assistance in many ways.

Acknowledgements

Preface

After the explosion in mobile telecoms back in the late 1990s with second-generation (2G) networks springing up all over the world, the race was on for the next generation of wireless communications systems. As 2G was and still is primarily used for voice, the way forward for communication traffic was assumed to be the predominant use of high-speed data transfer with the future third-generation (3G) networks. Although this is still yet to prove to be the final reality, the technologies in use have advanced at a phenomenal rate, therefore ensuring fundamental changes with both the network hardware and software. In addition, these new technologies have brought about vast changes within the network planning, implementation, and optimization of these future networks.

It is with these changes in mind that this book was originally conceived and developed, in order to specifically explain the fundamentals and the interacting elements involved in performing the next generation of network planning.

With 3G planning a whole different philosophy applies, the relative simplicity involved in the planning of 2G, due to the finite parameters, is now not applicable. The planning of a 3G network is now a complex balancing act between all the variables in order to achieve the optimal coverage, capacity, and quality of service (CCQ) simultaneously. This will not only prove challenging in the initial stages, but due to the nature of the anticipated services and their variability in demands on the system, this challenge will continue with the ongoing optimization of the network.

This book provides a detailed description of how the planning of a 3G radio network should be performed, along with the optimization techniques required to maintain a stable network. The basis of this book is structured around a CCQ model, which is continually referred to throughout this publication and should be considered the basis for successful 3G radio network planning.

This book is targeted at cellular operators, terminal manufacturers, radio and network planners, communication engineers, implementation managers, and all those interested in UMTS and the concept of network planning for 3G.

Introduction

1.1 General Mobile Telecoms

The universal mobile telephony system (UMTS) third-generation (3G) networks will ultimately provide convergence between mobile telephony, broadband access, and Internet protocol (IP) backbones. This will allow the mobile radio environment the extensive range of Internet applications and access speeds that are commonly experienced in fixed networks. The 3G mobile systems are based on a system known as code division multiple access (CDMA). This has fundamental differences to the current second-generation (2G) systems, which utilize the time division multiple access (TDMA) technology, widely used in networks throughout the world today. UMTS represents the 3G mobile telecommunications systems, which will, in the majority of cases, supersede 2G* and 2.5G** mobile systems.

1.1.1 Mobile Telecoms Today

Since the end of the technology and telecoms boom back in 2000, many operators now possess large debts; furthermore a rough estimate of mobile operators' capital expenditure on a global scale shows that it has collapsed from around $240 billion to $130 billion in one year (according to information from *BWCS Mobile Data Daily*, January 2003). In turn this has also caused the equipment vendors, such as Nokia and Ericsson, to reduce their capacity for implementation and equipment development.

* Such as global system for mobile communication (GSM), cdmaOne, and TDMA.
** Such as general packet radio systems (GPRS).

On the other hand, the current voice traffic on the GSM networks is still extremely profitable for existing operators. However, the market today is far more competitive, hence there is constant pressure on operators to reduce their charges, thus lowering profit margins. In addition, there is still the threat of substitution by low-priced voice over IP (VoIP) calls.

In Europe mobile penetration saturation is around 70–80 per cent of the population. Short message service (SMS) services are still continuing to generate solid revenues, however most other 'wireless Internet type' services have not generated anywhere near the expected incomes that the operators had hoped for. In addition, GPRS (2.5G) services have been very slow on the uptake, as today's customers are far more aware of the costs involved with regard to data services at present. Finally, another highly profitable area for most operators, the call termination charges, is likely to suffer due to regulatory pressure.

1.1.2 Development to 2G Status

First-generation mobile communications refer to the original analogue cellular systems that were soon superseded by the digital systems currently in use, known as the 2G systems. These comprise of the well-known GSM systems employed throughout most of the world today. In addition, cdmaOne (IS-95), personal digital communications (PDC) systems, and US-TDMA (IS-136) also fall under the 2G category, and are implemented in a few countries dotted around the globe including the USA.

1.2 Transition from 2G to 2.5G to 3G

3G refers to the IMT-2000 system, developed with the International Telecommunications Union (ITU), and is about to be implemented in many countries as universal mobile telecommunications system (UMTS). The US, European, and Japanese mobile operators' platforms differ slightly, however unification within a single standardization process will ensure that eventually the end user will have the ability to roam between different systems in different countries.

The advent of digital 2G systems has revolutionized mobile cellular communications worldwide, and GSM networks now exist in virtually every country with varying degrees of coverage. Subscriber levels based on the working population are now reaching saturation point and operators are now tending to focus more on their business and high revenue generating customers.

Additional services, such as limited data services and the immensely popular SMS system, have currently paved the way towards further requirements for a wider range of services. A new generation of cellular systems that can deliver both higher bandwidth and faster data rates has long been seen as the way forward, and subsequently the gap between fixed line data services and mobile cellular services will diminish further as more and more content becomes available.

1.3 3G Universal Mobile Telephony System

The ITU has certain minimum requirements for the data speeds that the IMT-2000 standards must support. These requirements will prove challenging to uphold as the physical speed of the terminal will have an effect on the data throughput of a 3G call. The data rate that will be available over 3G will depend upon the type of environment in which the terminal is located. For high mobility, it is expected that 144 kbps should be possible for outdoor mobile use. This data rate should be possible when the 3G user is travelling at speeds greater than 120 km/h in outdoor environments. For full mobility, it is expected that data rates up to 384 kbps could be attained for pedestrian users and those travelling at speeds less than 120 km/h in urban outdoor environments. Finally, for limited mobility, at speeds of less than 10 km/h in indoor and outdoor environments, data speeds of 2 Mbps will be possible. These potential maximum data rates that are often talked about when illustrating the potential for 3G technologies will only be available when utilizing certain technologies and under limited conditions, for a limited amount of users at any one given time.

Despite the media hype, in reality it is not likely that these high data speeds of 2 Mbps will materialize for some time to come. Infrastructure costs for the large and ever-expanding number of base stations that will be required are also significantly and potentially problematic, particularly due to increasing difficulties with site acquisition. This last factor, coupled with various technical issues that will affect each operator, will make it advantageous for operators to work together, thus sharing the costs. There remains a further number of technical issues, particularly the terminals, as well as growing public concern over the number of base stations required for operators to successfully provide the new range of services. The timing for some European operators in the purchase of the UMTS licences has been unfortunate, as due to the heightened market obsession with the technology sector, it appears that some operators have significantly overpaid for their licences. However, when examining whether operators have 'overspent' it is important to be objective with regard to the future of mobile telecommunications, especially the future role of data services. It is commonly predicted that the potential revenues that can be achieved from simple and complex data services will generate the majority of revenues.

1.3.1 Early Developments

In December of 1998 the 3G partnership project (3GPP) was born following an agreement between the major standard-setting bodies: the European Telecommunications Standards Institute (ETSI), the Association of Radio Industries and Businesses (ARIB) of Japan, and the American National Standards Institute (ANSI) of the USA. The benefits to be realized from this unprecedented co-operation of these bodies started the process towards creating a global standardization procedure. This process requires that all infrastructure vendors and mobile operators worldwide adhere to the agreed standards. Hence, the 3GPP is responsible for preparing, approving, and maintaining the technical specifications and reports for a 3G mobile system, based on evolved GSM core networks (CNs) and the frequency division multiple access (FDMA) and time division multiple access (TDMA) radio technology (see Figure 1.1).

Early in 1999, progress was made in agreeing a global IMT-2000 standard that met the political and commercial requirements of the various technologies available (such as GSM, CDMA, and TDMA). The 3GPP has now been divided into two organizations developing the approved standards, the 3GPP-1 and the 3GPP-2. The 3GPP-1 is responsible for defining and agreeing the global specifications for the GSM network evolution to UMTS, including the universal terrestrial radio access

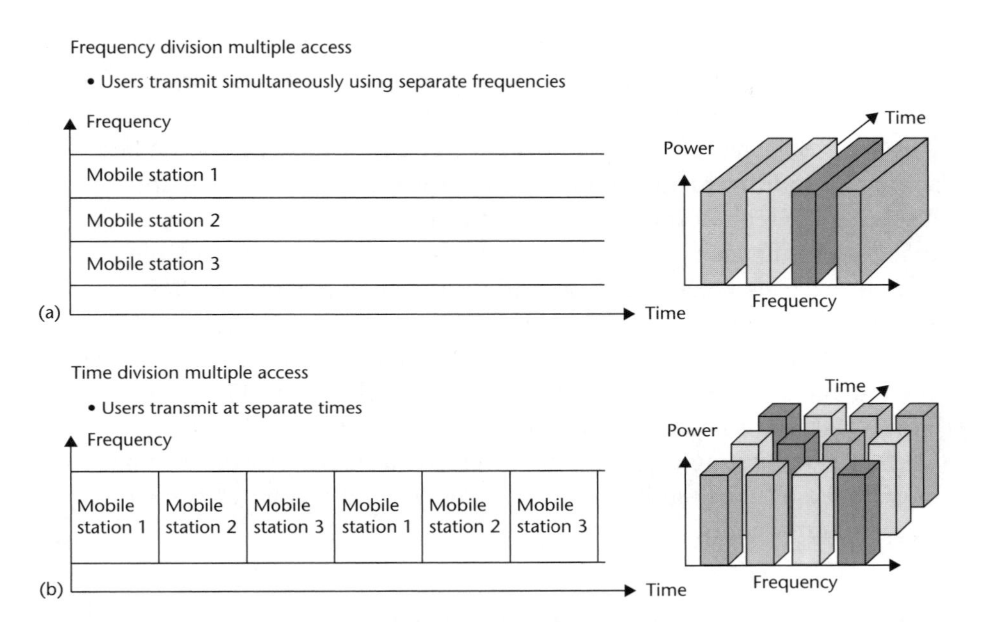

Figure 1.1 (a) FDMA and (b) TDMA.

network (UTRAN) and the CN for the wideband-CDMA (W-CDMA) variant throughout the majority of the world, including Europe and Japan.

The birth of the 3GPP-2 was mainly due to work performed within the Telecoms Technology Association (TTA) of Korea, and the 3GPP2 now maintains the standardization processes for the direct sequence (DS) and multi-carrier (MC) cdma2000 variant.

The interim step between 2G and 3G systems comprises of GPRS, and is often referred to as 2.5G. GPRS is considered the 'stepping stone' towards the long awaited 3G UMTS systems, which will offer the rich multimedia services of the future. However, the success of GPRS currently implemented by various operators is being carefully watched, since to a large extent this will influence both the already jittery markets and the media as to whether 3G will potentially be a success. Ironically, just as with GPRS, numerous technical problems still have to be ironed out, along with software problems within the already limited numbers of terminals currently available. The over-hype by the media has tainted the general public's perception of the actual data speeds available, and the promised

2 Mbps data connections along with seamless video-to-video services that have not materialized, have not helped. Predicted data usage in the past has for the most part been widely inaccurate. Generally speaking, it appears that the relatively slow uptake of GPRS subscribers can be attributed to introducing and marketing the service too early before resolving the technical issues with both the network and the terminals. In addition, the lack of available terminals and the lack of available content have also dampened the enthusiasm with regard to GPRS. A valuable lesson can be learnt here to ensure the future success of 3G UMTS cellular systems. Finally, due to the inherent differences between 2G and 3G, accuracy of the radio planning for the 3G UMTS systems is a critical factor, as will become evident in the following chapters. Correct 3G planning is paramount to ensure a stable network can be implemented fulfilling the capacity, coverage, and quality-of-service (QoS) requirements.

1.3.2 Universal System Agreements

In 1998, the ITU (see www.itu.int) requested radio transmission technology (RTT) proposals for IMT-2000 (originally called future public land mobile telecommunications systems), the formal name for the 3G standard. Various different proposals were submitted. The digitally enhanced cordless telecommunications (DECT) system and TDMA/universal wireless communications organizations submitted plans for the RTT to be TDMA based, however, all remaining proposals for non-satellite-based solutions were based on W-CDMA. The main proposals were called W-CDMA and cdma2000. The ETSI and the GSM operators including the infrastructure vendors, such as Ericsson and Nokia backed the W-CDMA proposal. In North America, the CDMA community led by the CDMA Development Group (CDG), including vendors, such as Lucent Technologies and Qualcomm, decided on the cdma2000 variant.

1.3.3 System Capability

As the existing profitable 2G systems will still continue to be utilized, the evolutionary migration to 3G systems will ideally consist of overlaying the new technologies, and running 2.5G (GPRS) and 3G alongside or in conjunction with the existing GSM networks. The basis of GPRS consists of overlaying a packet-based technology on top of the existing circuit-switched GSM network. Packet-based switching networks have the ability to move data in separate, small blocks known as packets. Within each packet exists the destination address. Once the sent packets are received, they are re-assembled in the proper sequence thus providing the required message. Circuit-switched networks require dedicated point-to-point connections during calls. This is considered a major step forward supplementing a circuit-switched network with packet-switched services. GPRS is able to provide the user with a greater increase in data throughput, and uses the existing radio resources more effectively. Fundamental changes are required within the CN, coupled with additional new network infrastructure.

It will also be possible to utilize another relatively new radio access technology in conjunction with GPRS known as enhanced data rates for GSM evolution (EDGE). EDGE will further enhance data rates over the air interfaces, which are backward compatible with GSM. EDGE is able to support both circuit- and packet-switched traffic, and the radio access is based on TDD.

1.3.4 3G Development Process

The first 3G network model to be widely implemented will be based on the 3GPP Release 99 specifications as illustrated in Figure 1.2. As can be seen there is a connection via the Iu, A, and Gb interfaces, between the UTRAN to the CN circuit-switched (CN-CS) domain. This is because the required quality of service for high data rate services cannot yet be provided via the packet-switched domain. In the fullness of time, as this is resolved, all traffic will eventually become packet switched. Packet is more advantageous, as it is possible to transmit and receive using different bit rates and accommodate multi-process communications. With a packet-switched network a connection cannot be refused as in a circuit-switched network, it is merely delayed until the packet can be transmitted. In short, network capacity is much greater and more efficient than with a circuit-switched system.

The business model will fundamentally change here, as the operator can now be thought of as a 'carrier provider'. An evolved intelligent network (IN) system will enable subscriber-based IN services to be transferred between networks. Within 3G the transmission medium will consist of

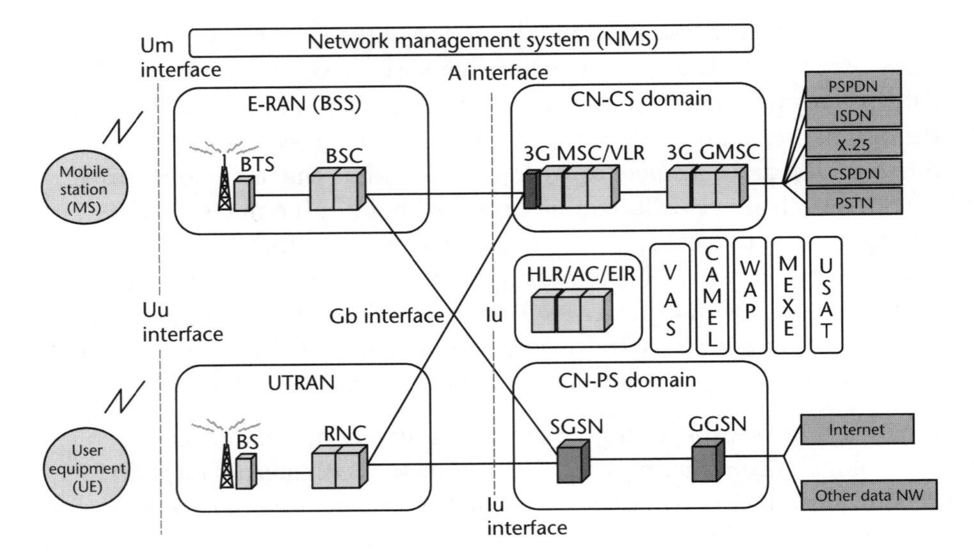

Figure 1.2 3GPP Release 99.

asynchronous transfer mode (ATM), which is well suited with regard to packet data transmission, as it is made up of very small cell sizes.

The next stage in the evolutionary path will be towards the 3GPP Release 4 (2000) as illustrated in Figure 1.3. Most of the changes in Release 4 will occur in the CN. The media gateway (MGW) is seen as a transport termination point between the 3G 'all IP' network and the public-switched telephone network (PSTN). All connections and switching functions are controlled by the master switching centre (MSC) server. A single MSC server will be able to handle numerous MGWs, and hence, the CN-CS domain is freely scaleable. The IP multimedia sub-system (IMS) will have the ability to offer more uniform methods to execute VoIP calls. Hence, it will be possible for a conventional circuit-switched GSM call to be converted to VoIP within the CN.

With regard to Release 5 (2000), as illustrated in Figure 1.4, the evolution continues further whereby all traffic from the UTRAN should be IP based. The major changes to be noted here are within the radio access part of the network, and also within the transport technology whereby ATM can be replaced by IP. In this scenario the operator will have the choice available whether to use IP or ATM, or even a combination of both.

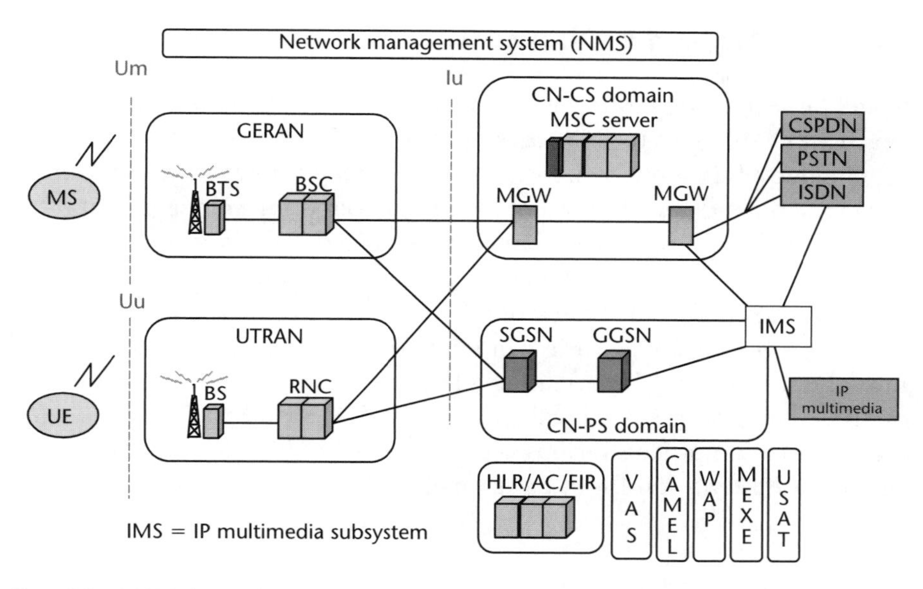

Figure 1.3 3GPP Release 4 (2000).

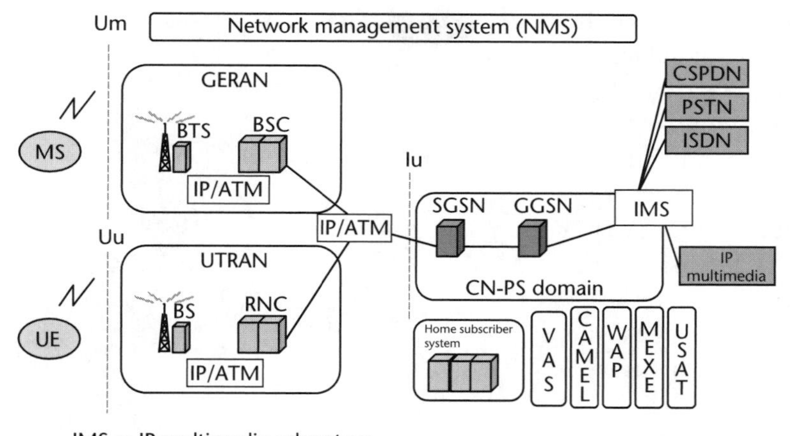

Figure 1.4 3GPP Release 5 (2000).

As the quality-of-service (QoS) issues are finally resolved and are fully functional in IP, the emphasis will be to move to an 'all IP' scenario. The ideal situation here is also to ensure that the CN has the ability to utilize several radio access technologies.

1.4 3G Network Planning

As will be seen throughout this book, planning of 3G networks will be considerably different to GSM-based networks. There exists tried and tested models for circuit-switched traffic behaviour, however the design, development, and construction of planning models for mobile packet-switched traffic is somewhat more complex. Needless to say such models are essential for both network planning and management. In addition, the network planner this time around is required to be far more knowledgeable on a wide variety of inter-related issues that are covered throughout this book. Ascertaining all the necessary information required for all planning works is critical and somewhat demanding. This information, comprising various estimations, will ultimately provide the basis for the network roll-out decisions and the predicted costs.

1.4.1 3G Roll-out Strategy

As illustrated in Figure 1.5 a UMTS roll-out will not equate to a standard GSM roll-out, whereby 3G coverage would encompass an entire country. It is neither physically practical, nor financially feasible to implement such a system. Only areas where high data traffic usage is predicted, such as densely populated areas, will be covered at present. The need for multi-mode terminals must not be forgotten, as this will be paramount in offering both 2G and 3G services simultaneously. With this

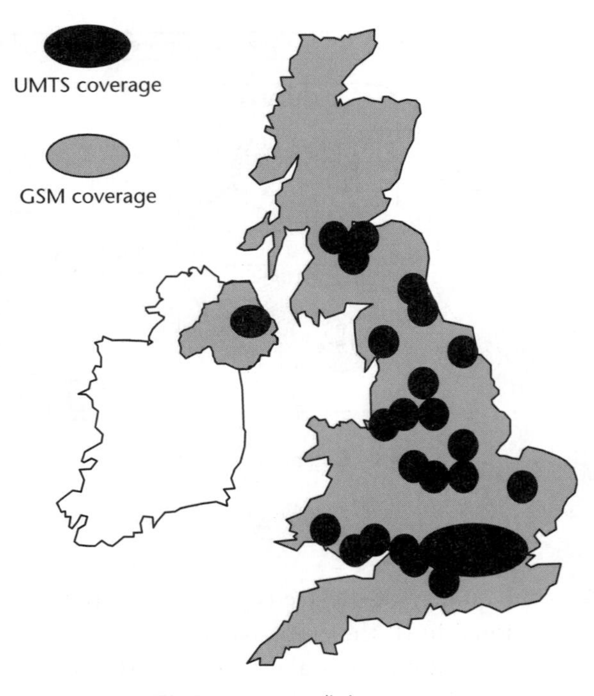

Figure 1.5 UMTS UK coverage prediction.

in mind, as covered in Chapter 5, it is evident that most existing operators will attempt to re-use the majority of their existing GSM sites.

1.4.2 The Coverage, Capacity, and Quality Model

The fundamental model on which this book is based can be seen in Figure 1.6. The coverage, capacity and QoS (CCQ) model illustrates the trade-offs that occur between these three critical parameters. This is the basis of quality 3G planning. Striking the right balance between these three parameters will have a significant impact on the ability of the network to perform in an optimal manner, and in turn will ensure the all-important return on investment (ROI) can be achieved. If it is performed correctly the business model will correspond to the real-life scenario. If the planning proves to be incorrect, then additional base stations will be needed, and the business case will be affected. Once these three trade-offs have been optimized, an individual balance can be achieved to provide the best 'real-life' scenario.

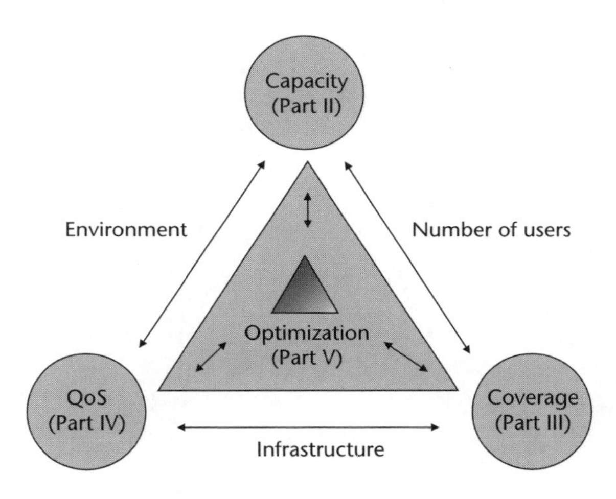

Figure 1.6 The CCQ model.

PART ONE

Network Planning and 3G Foundations

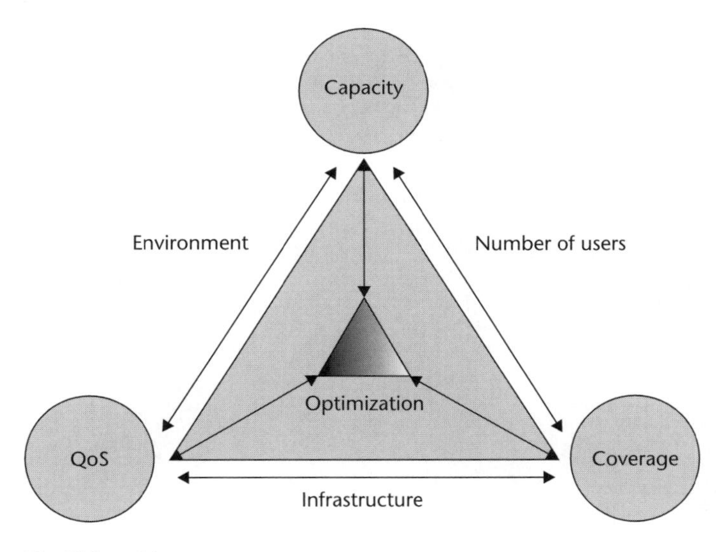

The CCQ model.

Introduction

This book is split into six parts covering all aspects of universal mobile telephony system (UMTS) network planning and introduces the coverage, capacity and quality-of-service model, known here as the CCQ model, which consists of the basis for UMTS network planning. In this part the fundamental differences can be seen between the old second-generation (2G) planning and the requirements for the next generation of UMTS (third-generation (3G)) planning.

In this publication the radio network planning process is based on the CCQ model and is referred to throughout each section in the book. The target coverage requirements, traffic densities, and possible coverage enhancements are discussed in this part, along with other important issues which must also be considered consisting of macrocell, microcell, and indoor coverage, co-locating with 2G networks and migratory traffic. This part is followed by Part II, which covers the network capacity issues.

Wideband-CDMA Network Planning

Designing a universal mobile telephony system (UMTS) network is a multi-dimensional process due to the large number of different design requirements and system parameters. In addition, when the number of users increases, the interference in the coverage area also increases, thus causing the cell size to shrink. Thus, third-generation (3G) network planning is a complex and challenging task. To ensure accurate 3G network planning, good estimates and forecasts of user traffic are therefore required. The topography, economics, and the logistics of site construction all have to be considered. An important aspect affecting the planning, roll-out speed, and coverage of a potential network is the co-location scenarios for both the existing second-generation (2G) sites, and other proposed 3G sites belonging to other operators.

When initially assessing the detailed plan, capacity and coverage are the two main issues that have to be considered. Within a wideband-code division multiple access (W-CDMA) network, the soft handovers will have an effect on the cell capacity and must also be taken into consideration (this is covered in more detail in Chapter 3 entitled 'Detailed Network Planning'). Once the network becomes operational, the required parameters, such as traffic statistics, types of services used, quality-of-service (QoS) statistics, cell/sector congestion, 'busy hour' traffic, and migratory traffic will all be downloaded to the network operation and maintenance centre (NOMC). The entire network can be controlled by the NOMC, as it is the heart and 'mission control' of the network. This control centre can also be known as the operations and maintenance centre (OMC), or the network operations centre (NOC). The propagation model tuning and re-planning/optimization will then form a continual process, which will be heavily influenced by the results of the downloaded parameters mentioned above. Once the network has been rolled out and is fully operational the traffic load will increase and it will be these traffic statistics that can be continually analysed. This will ensure that network can continually be re-optimized and re-planned to match the traffic requirements as illustrated in Figure 2.1.

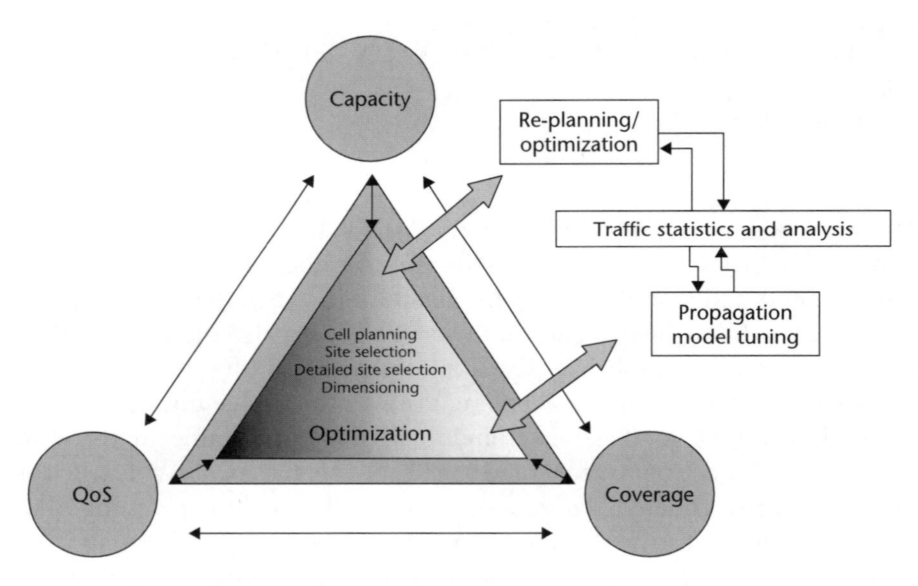

Figure 2.1 Processes required in network planning and roll-out.

The basis of W-CDMA planning consists of the development and establishment of the planning process and the network optimization procedures, followed by the development of the actual network design plan. The final stage involves the optimization and re-planning, which will be an ongoing iterative process.

2.1 Radio Network Planning Process

Firstly, it is necessary to estimate the number of cell sites, the type of base stations and their configurations (including the number of network elements, sectors, and antenna configurations). In order to attain the number of cell sites and their configurations we need to assess the coverage, capacity, and QoS (CCQ) requirements together with the type of area to be covered, such as a dense urban area. With this information available it will be possible to start the dimensioning, coverage, and capacity planning. The output from this coverage and capacity planning, along with the dimensioning will provide an approximate number of sites and their configurations.

To acquire an understanding of the proposed requirements the following three factors must be determined:

a) Capacity;

b) Coverage;

c) Quality.

As with so many aspects of the W-CDMA system, these factors interrelate with each other. There is a fourth controlling factor that is time dependent and will have a bearing on these three fundamental criteria in the long term. This fourth factor is known as the optimization process and is discussed in Part V, entitled Optimization and Network Planning, of this book.

Within a 3G network, the planning and optimization/re-planning tasks will become an ongoing process. Once the capacity and coverage planning has been achieved and a 'useable' network has been built, the terminals will then be introduced into the network in a controlled fashion allowing further network performance and visualization. This ongoing process will ensure that performance will continually be monitored for optimum efficiency.

The main objective of network planning therefore is to efficiently control and maximize the effects of these four parameters.

2.1.1 Capacity

W-CDMA capacity is a somewhat multifaceted issue and is dependent on numerous variable factors which interact with each other. For example, as more users enter a cell and their data rates increase, this will diminish the cell's capacity. In addition, if the cell load increases then the interference levels will increase; this also inhibits the cell's capacity. Finally, soft capacity is another issue to be considered, because in certain situations some cells may be less loaded than others, and hence this excess capacity can be shared between cells.

2.1.1.1 Capacity Boosting Techniques

Some useful capacity enhancing techniques can be supported within a W-CDMA network. Methods used can consist of macro, micro, and pico cells as overlays or underlays in a cellular hierarchy. Using these three types of cells together in this manner is known as creating a hierarchical cell structure (HCS). The HCS is a classic method for increasing capacity in a hot-spot. The main idea is to use different frequency bands for different kinds of users; however, this assumes that the operator has purchased enough of the available frequency spectrum to take advantage of HCS. For example, an operator can purchase three separate frequency 'blocks' of which each will consist of three blocks for the uplink and three corresponding blocks for the downlink. This operator will have a greater opportunity to 'switch' to an alternative frequency in a dense urban environment, thus reducing the possibility of any adjacent channel interference (ACI), or alternative operator interference.

Spatial division multiple access (SDMA) is implemented by organizing individual antenna elements in an array. This mechanism can be used to increase capacity and reduce interference. SDMA enables techniques that can continually monitor the required coverage areas and enjoy the ability to adapt to the ever-changing radio environments. The system provides an effective antenna pattern that is able to adapt its radio frequency (RF) propagation pattern to 'follow' and 'track' the movements of the user. This technique, SDMA, can be likened to our ears. We can locate the source of a sound because we have two receivers separated by a small distance. Both ears receive the sound at different times depending on the direction of the sound. The brain then processes this information from both ears to determine the direction a particular sound is coming from. Further benefits are provided by SDMA in the enabling of range extension, interference reduction and, in certain cases, using multi-path signals (copies of the desired signal) to provide a reinforced signal.

Multi-beam (or adaptive) antennas are often referred to as smart antennas without handovers between beams. With adaptive array antennas, the received signals from the multiple antennas are combined and weighted (brought into line with each other) to maximize the signal-to-noise ratio (SNR). We use a 'RAKE' receiver to allow each antenna to track a signal and compensate for fading. In this case 'RAKE' is not an acronym, the name 'RAKE' was coined by its inventors Price and Green in 1958. When a wideband signal is received over a multi-path channel, the multiple delays appear at the receiver. By attaching a 'handle' to the plot of the multi-path returns, the visual impression is that of an ordinary garden rake. It is from this picture that the rake receiver gets its name. The best way to describe a RAKE receiver is that its function is to sample, compute, and reconstruct the damaged multi-path signal rays; so the final result is a reconstructed and reinforced signal. The RAKE receiver resolves and combines the desired signals for one user, which have been received via a number of different propagation paths. A RAKE receiver consists of multiple 'fingers' which can be thought of as separate receivers. Each of these 'fingers' de-spreads a desired signal and the relative time offsets are then adjusted out (eliminated) with delay circuitry. The end result is that all signals from each 'finger' are added together, thus achieving a reconstructed and reinforced signal.

2.1.1.2 Soft Capacity

Soft handovers employ multiple connections to multiple base stations, all of which operate on the same radio frequency (RF). The use of multiple connections reduces the power from each base station serving the user equipment (UE). Soft handovers will thus result in a reduction in interference and in turn will lead to an increase in cell capacity. Soft handovers can therefore provide the soft capacity within a cell. This is an important issue that must be taken into consideration by the planner and as such is covered in further detail in Chapter 3, entitled Detailed Network Planning.

2.1.2 Coverage

Coverage within 3G will ultimately enable a user to roam seamlessly from the home network into other 3G networks. In addition, coverage will be provided in a hierarchical type of cell structure within the network whereby users will hand off from a picocellular/microcellular environment to a macrocellular environment. A UMTS network will not initially mirror the existing global system for mobile communication (GSM) coverage and hence there will be 'islands' of UMTS coverage, which will co-exist with the existing GSM systems. Multi-mode terminals, which will be able to switch from 2G to 3G and vice versa, will therefore be required.

2.1.2.1 Target Coverage

The variety of different services available will demand different levels of coverage due to factors such as the different energy-per-bit/noise-per-spectral density (Eb/No) (discussed further in Section 6.2.4), required to obtain acceptable levels of service, different processing gains and soft handover scenarios during which no soft handovers will occur. This type of scenario will occur within a HCS whereby one cell is overlaid on top of another, yet both cells are employing two different frequencies; hence, soft handovers will not be possible. Finally any other link budget variances, such as base station noise figures, receiver noise power, etc., will have a universal influence on the factors of CCQ. All of these are covered in more detail in Chapter 6, entitled Influence of Link Budgets on 3G Coverage.

The target coverage must be taken into account including the environment where a user is located, along with various range calculation parameters (these are also covered in detail in Chapter 6). Coverage polygons can be used to depict roughly the areas where specified coverage is required. These can also act as a basis for acceptance testing at a later date and will therefore be beneficial to the operator. It is not advisable to create coverage polygons specifically from data depicting the types of clutter present (clutter is further explained in Chapter 2). Coverage levels for dense urban, urban, and suburban areas will generally form concentric rings around a town. It is not advisable to attempt to create a cell plan whereby overlapping between towns occurs. For road coverage it is more beneficial to use a list or a schedule of roads to be covered, instead of creating polygons that would link towns. This would result in rather unconventional shapes that could easily overlap with each other. As can be seen in Table 2.1, for dense urban coverage assuming a target data rate of 128 kilobits per second (kbps) a coverage availability of 95 per cent could be achieved, as opposed to a suburban area, assuming the same data rate of 128 kbps, whereby only a coverage availability of 90 per cent can be realized. The types of areas depicted in Table 2.1 are covered in more detail in Chapter 9, Radio Environments and Microcell Planning.

Type of coverage	Type of environment	Target service rate (kbps)	Type of service	Coverage availability (%)
Dense urban	In commercial buildings	128	PS	95
Urban	In commercial buildings	128	PS	90
Suburban	In residential buildings	128	PS	90
Road/Rail	In car	64	PS	90

Table 2.1 Target coverage.

2.1.2.2 Sectorization and Adaptive Beam Forming

Sectorization consists of splitting a cell into sectors, which in turn will ensure that an increase in capacity can be achieved. In addition, to enhance the coverage area of these sectors it is possible to use adaptive beam-forming antennas, which have the capability to 'track' the UE as it is moving, thus increasing the actual coverage area. When considering cell sectorization or adaptive antenna beam forming, estimates can be made to ascertain the spectrum efficiency requirements these techniques can address. These will be dependent on the specific type of radio environment and the antenna equipment. In an open space environment, for example, the maximum sectorization will provide a somewhat higher downlink gain than in a street-level-type microcell environment.

2.1.3 Quality of Service

Within 3G networks, QoS is paramount as compared to the previous 2G systems. This is due to the variety of high data rate services that will be available over a wide bandwidth. These services include facilities such as video-to-video and data exchange which will require high levels of QoS, thus ensuring that adequate transmission and reception can be attained. Therefore as with capacity and coverage, QoS is another important issue to be carefully considered within a UMTS system, due to the nature of the services and end user requirements.

There are different levels of QoS, which enable a wide range of applications to be supported. At present it is still unclear what the nature and usage of some of these new non-voice services will be as they are still evolving, so it is not logical to optimize UMTS for only one set of applications.

2.1.3.1 Planning with Regard to Quality of Service

It is more difficult to predict the different types of traffic and the expected usage of different services such as video-to-video and e-mail than to plan for the provisioning of circuits. Video requires high data usage in real time; it is similar to providing for circuit-switched voice but with much greater bandwidth demands. E-mail, in comparison, requires low data rates, which could be performed with enormous delays in non-real time.

Traffic density must be considered when evaluating aspects of network quality, such as blocking and outages. The changing numbers of users in a cell, each with their own changing QoS demands, create a challenge to achieving uniform coverage over a wide area. It is also important to assess as accurately as possible what effect different cell deployment techniques will have on the usage and requirements of a wide array of predicted services due to evolve in the very near future.

Finally, co-location and co-existence with an existing 2G GSM system must be taken into account, as there are numerous issues the planner must be aware of with regard to the implementation of a W-CDMA system. These are reviewed in Chapter 5, 3G Co-planning and Co-existence.

2.2 Developing Procedures for Planning and Maximizing Capacity and Coverage

The CCQ parameters must be considered to ensure that effective procedures can be developed for the network plan, as each of the three parameters, coverage, capacity, and QoS must be taken into equal consideration, thus ensuring an optimal balance between these three parameters can be achieved.

The W-CDMA air interface is unpredictable and constantly changing. This is due to the variable amount of users and their changing data rates. Each cell must be carefully planned and hence the prerequisite to achieving an accurate network plan includes the following three phases:

a) Preparation;

b) Number of estimated cells;

c) Management of the network plan.

2.2.1 Preparation

The preparation phase consists of evaluating the initial capacity and coverage requirements, defining the network planning strategy, verifying the operating constraints, and finally preparing the preliminary design. These depend on the operator's specific requirements and so at this stage these will be based on prediction inputs obtained from the marketing exercise.

The initial consideration for the preparation is the number of users to be served. For this reason the following issues are considered when calculating the number of users requiring service:

a) population working in a specific area;
b) population living in a specific area;
c) vehicular traffic through or in the area;
d) recreational areas and events (e.g. concerts, sports events, etc.).

The offered traffic per user will depend on what types of services are available and how frequently they are being used; these services will also be somewhat fluid as many are still yet to be introduced. The average calling time for circuit-switched users will naturally depend on the type of service in use (e.g. video, speech, etc.). The above will also be dependent on different customer types, which can be split into two main sectors: business/corporate users and private users. As an example, typical values for the average calling time for speech services can range from around 120 to 180 seconds. It will be somewhat more difficult due to the wide variety of known and future UMTS (3G) services to calculate an average calling time where calling times currently apply as most of the new services will probably be packet switched. Combine this with the fact that some services, notably packet services, do not involve connection or calling times at all as they are considered to be permanently connected. However, since the average calling time has a significant effect on the generated network traffic, predicting this with some degree of reasonable accuracy must be considered a critical factor. This is made even more difficult due to the added new dimension to UMTS (3G) services, that of an average volume of data throughput, which at the moment is also unknown. To complicate the issue even further, the demand for certain services (e.g. coverage of sporting events) must also be taken into account, as this will vary according to the time of day.

2.2.2 Number of Users

Calculating the number of users in a specific area is achieved by multiplying the penetration by the population. The number of users and the offered traffic for each user yields the offered traffic.

2.2.3 Management of the Network Plan

Once the number of cells has been determined, detailed radio planning can be performed, which will consist of determining how the RF will propagate in each required cell. This will influence the initial coverage as building restrictions, costs, construction planning laws, site configurations, co-location with other operators, and ACI must also be considered. For these reasons a professional network planning tool will be required due to the near impossibility of being able to acquire and/or construct sites in the optimum locations.

There are numerous tools available on the market; a few examples are those offered by such manufacturers as ATDI, Logica, Commweb, and Agilent. A superior network planning tool will also include a suite of modular software 'add-ons' which can be integrated into the existing planning application. Modules can include information management tools, cell site design tools, equipment configuration, inventory tools, network performance monitoring and reporting tools, and project or general management tools. These can be purchased as and when required, giving the operator more flexibility depending, for example, on which aspects of planning must be performed in-house. The more third party services used the fewer modules that need to be purchased.

To summarize, it is important to be aware of the three stages, namely preparation, estimation of number of cells, and finally the management of the network plan, which are required to ensure correct procedures can be developed, thus enabling the maximum coverage and capacity to be achieved for the proposed network.

2.3 Development of the Network Design Plan

When considering a high level view of network planning design, the following subjects must all be considered and are all covered in detail in the following chapters:

a) Microcell/macrocell coverage (Chapters 3 and 5);
b) Microcell/macrocell capacity (Chapter 3);
c) Indoor and high data rate coverage (Chapter 9);
d) Co-location (Chapter 5);
e) Eventual migration from GSM networks (Chapter 5).

The policy adopted for network deployment will be heavily dependent on the actual traffic distribution. A practical approach would be to use macrocells for outdoor coverage and picocells for indoor office coverage. Macrocells can also fill gaps in indoor coverage. However, in such a case, it would be beneficial to provide extra capacity by increasing the number of indoor cells, as opposed to providing indoor coverage from outdoor cells. If an increase in indoor cells is not implemented, it could lead to a scenario where the operator might be forced to introduce microcells to cover the capacity restrictions. It is possible, however, to have high data rates uniformly over the whole cell area, or alternatively the data rates available could be lower when the user is at the periphery of the cell, which would therefore enable a larger cell coverage area to be achieved. This will depend on the type of high data rate services that may be available, for example video-to-video. For services that require a lower QoS profile, non-uniform coverage could be considered, such as e-mail and store and forward services such as Short Message Service (SMS). However, for real-time services requiring a high level of QoS, uniform coverage will be needed. QoS is covered in more depth later in Chapter 7. Finally, if an operator has an existing 2G network, user transfer characteristics must be considered, such as inter-system handovers, cell re-use, and co-existence with external operators.

Offered traffic per user is heavily dependent on the types of available services, and how frequently they are used. The cell capacity estimates can be based on simulations or analytical formulas. To estimate the maximum cell coverage, link budgets must be taken into account (this will be covered in Chapter 6). In addition, the Eb/No (pronounced 'ebno') performance, cable loss, antenna gain, receiver noise figure, and soft handovers are also covered in Chapter 6. Finally, the number of estimated cells in the network can be finally calculated based on the capacity and the link budget.

2.3.1 Traffic Analysis

Traffic analysis consists of the following:

a) Examining the demographic spread throughout the coverage area;
b) Market projections to identify subscriber distribution;
c) Calculating offered traffic per subscriber;
d) Creating a traffic 'map' indicating the amount of expected traffic throughput.

This should be a task for the marketing department, however, the network planners will require traffic maps to enable network dimensioning to be performed. This will also allow an initial network capacity analysis of the nominal plan to be completed, and to simulate the performance of the final plan.

2.3.1.1 Traffic Intensity

The evaluation of both the traffic intensity (the offered traffic) and the actual user traffic in a specific area must be determined. The data delivery density (DDD) is defined as the information rate that can be delivered per coverage area. First, we have to establish a unit of measure that quantifies traffic intensity.

$$\frac{\text{Call time}}{\text{1 hour (h)}} = \text{Quantifiable traffic} \tag{2.1}$$

For example, if a subscriber makes two calls in 1 h and the average length of these calls is 120 seconds (s), the above formula gives the following:

$$\frac{240}{3600} = \text{66 milli Erlangs (mErl) of traffic}$$

Two calls per hour at an average of 120 s equals 240 s, and there are 3600 s in 1 h. Erlangs (Erl) were designed as a quantifiable measurement of traffic and can be thought of as the continuous use of one voice path. In reality, it is used to describe the total traffic volume in 1 hour (h). There are several Erlang models available, however, the most commonly used model is the Erlang-B, which can be used to evaluate how many lines are required, assuming the traffic figure (in Erlangs) during the busy hour is known. With the Erlang-B model it is assumed that all blocked calls are immediately cleared. If the user density in a region is considered, then the traffic density can be calculated and is expressed in Erlang per square kilometre (Erl/km^2). A single measurement for traffic may not prove appropriate for all cases, especially since there is a diverse assortment of services in UMTS (3G) networks. For data services, the volume of traffic is better characterized by performing the measurements in megabits per second per square kilometre (Mbps/km^2) which is the capacity to transmit data in a specific time period over a specific area.

2.3.1.2 Sample Usage and Service Offerings

Market penetration, service offerings and service usage per user must be considered. Business market penetration consists of the number of subscribers per head of the population divided by the number of business employees. This may then be broken down into the percentage of the penetration for the different target groups (e.g. the business user). In addition the service offerings include the types of services offered and the data rates they demand including their QoS profiles. For service usage, estimates must be made for circuit-switched services in Erlangs or calls per busy hour and for packet-switched services in megabits per hour (Mbph) per busy hour. Traffic analysis must account for a variety of services (e.g. voice, data, video, etc.) and also various environments, such as indoor and outdoor, along with private and public usage. A traffic model will be required using a set of simplifying assumptions, for example, the distribution of certain random time periods (e.g. call duration and 'cell usage time'). This type of traffic model should allow for the duration of the call and the estimation of the handover rate (covered further in Section 8.7). This will require an iterative method to enable the calculation of the offered traffic load per cell. In addition, another parameter required is an estimate of the time it takes a busy mobile user to leave a cell area and an estimate of the cell periphery crossing rate. This will enable a more accurate prediction of user traffic within the specified coverage area.

From both the market penetration and the demographic/business data, the number of users in a specific area can be calculated.

To create a subscriber map suitable for network capacity simulation or analysis, different land usage categories must be considered. If the service usage is then applied per subscriber, a suitable traffic map can be created. This could be used to assist in network dimensioning, or even imported into the planning tool. The differences in data usage can be seen in Table 2.2 from the high interactive multimedia, such as video-to-video, to the low volume data, which could consist of services such as e-mail delivery. For example, Table 2.2 shows that for high interactive multimedia-type services the data usage is likely to be around 128 kbps, thus an average call length for this type of connection will be around 144 seconds. In contrast, for a speech call using around 12.2 kbps the effective call duration is likely to be around one minute.

Service	Usage kbps	Average effective cell duration (s)	Switch mode (circuit of packet switched)	Calls/subscriber per busy hour
High interactive multimedia	128	144	CS	0.25
High multimedia	2000	53	PS	0.01
Medium multimedia	384	14	PS	0.2
Low volume data	14.4	156	CS	0.1
SMS	14.4	30	PS	2
Speech	12.2	60	CS	1.5

Table 2.2 Sample usage and service offerings.

2.3.2 Cell Coverage Estimation

To obtain more detailed coverage predictions, some correction factors must be added for the intended required pathloss models. Field measurements can also be used where possible. The pathloss models are calculated using a combination of empirical data and analytical techniques. These can only be confirmed by comparing the pathloss model with the data acquired from the actual frequency and the precise coverage

area required. These models can be used for system simulation.
For the uplink, the impact of load factor 'n' in the link budget, and the interference margin 'I_m' (in dB), can be calculated from the following formula:

$$I_m = 10 \log\left(\frac{1}{1-n}\right) \tag{2.2}$$

As illustrated in Figure 2.2, the cell range will decrease with an increasing load factor as the interference margin increases. Multi-user detection (MUD) can reduce the effect of the load factor on the range of the cell. MUD is also sometimes known as joint detection and enables interference cancellation, thus reducing the effect of multiple access interference (MUD is discussed further in Chapter 4). A higher capacity exists when using MUD, as most of the interference is intra-cell interference. If MUD is implemented, then users in a soft handover scenario could experience an extra 20 per cent of interference to be cancelled out in a macrocell environment. The cell separation is somewhat better in microcells, so fewer users are in a soft handover scenario and thus the levels of interference from other cells will be less than experienced in macrocells.

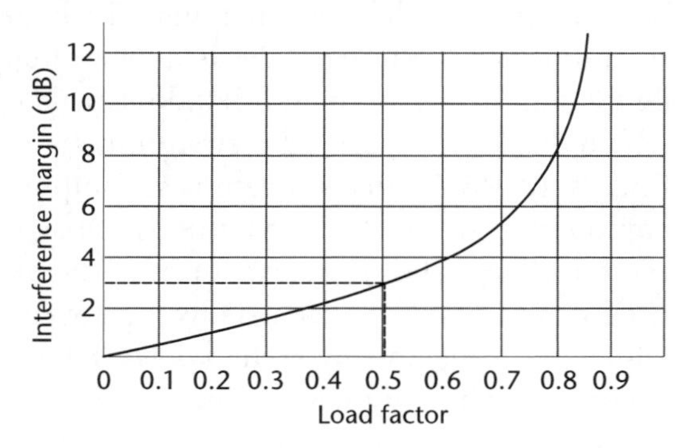

Figure 2.2 Maximum cell coverage estimation – load factor.

Asymmetric traffic must also be taken into account when attempting to calculate link budgets. It is possible to trade the uplink system load for coverage within a W-CDMA system, due to the user equipments (UEs) power actually limiting the maximum cell range, which in turn is beneficial.

2.3.2.1 Link Budget

Link budgets, which are discussed in detail in Chapter 6, are calculated separately for the uplink and downlink. Antenna gain, receiver noise figure, and cable losses also contribute to the calculations.

2.3.2.2 Soft Handover Gain

The soft handover gain has a distinct effect on the link budget. The soft handover gain is brought about by handoffs at the boundary between two or more cells and where there is an equal average loss to each of the participating cells. To obtain the soft handover gain, the log-normal fade margin must be calculated. This is the margin required to provide the specified coverage probability at the periphery of a single isolated cell and then the corresponding margin required at the periphery between two or more cells. Therefore, the soft handover gain is the actual difference in dB between the two different margins. The larger the coverage probability requirement, or alternatively the lower outage probability, the greater will be the required margin. Soft handovers also provide macro-diversity gain, which depends on the radio environment and the number of RAKE fingers. (A RAKE receiver samples, computes and reconstructs the components of the damaged multi-path signals, yielding a combined and reinforced signal of the correct phase and amplitude.) A RAKE receiver used in W-CDMA service usually has four to six 'receivers' (fingers), which are adjustable by system delay. Each finger consists of a local code generator, gain and phase tuning equipment. Further information can be found on RAKE receivers in Section 2.1.1.1, Capacity Boosting Techniques. However, soft handover gain should always be taken into consideration as this can assist in improving the link budget.

2.3.2.3 Coverage Limited

A network is 'coverage limited' when it has sufficient capacity to handle all of its traffic, but from a capacity viewpoint, the cell size (or coverage) could be greater. The maximum cell range is the limiting factor. The maximum cell (or coverage) area is used to determine the number of base stations required.

2.3.2.4 Capacity Limited

A 'capacity limited' network is one that cannot support all of the offered traffic. The cell count is found by dividing the number of users one cell can support by the number of users per square kilometre. This is the same as the cell area in square kilometres. If the total area is divided by the cell area, from this it is possible to ascertain the total number of cells the system requires. With regard to 'interference limited' networks, a smaller signal outage probability will result in smaller cells, and thus increased costs to implement the extra cells required (this is covered in more detail later in this chapter). Alternatively, a smaller interference outage probability will result in less capacity being available. This presents a similar problem, as further costs will be required to increase the capacity. It is important to remember a trade-off exists here and hence an attempt should be made to achieve a stable state between both these extremes.

2.3.2.5 Load Factor

The system has reached its pole capacity when it is loaded at 100 per cent capacity, or when the maximum theoretical capacity has been achieved. It can be stated that the pole capacity of a cell is dependent on the processing gain, the average Eb/No target and the voice activity factor. The coverage area, whereby all users obtain the target Eb/No, depends on the loading relative to the pole capacity. At pole capacity the cell size would shrink to zero and the rise in interference levels would approach infinity. Pole capacity is directly related to the load factor and a UMTS (3G) network cannot be operated at pole capacity. To enable a suitable load factor to be determined, a number of issues need to be examined which are discussed further in Chapter 4.

Load Factor Example

A 64 kbps data service has been assumed. Based on a network simulation program, we conclude that with a full load the spectrum efficiency is 50 kbps/MHz/cell (e.g. 250 kbps/cell, given a 5 MHz bandwidth). If a 50 per cent load is assumed, then the actual spectrum efficiency would be 125 kbps/cell.

This is derived from the following: As spectrum efficiency at an assumed 100 per cent cell load is 250 kbps/cell, therefore spectrum efficiency at a 50 per cent cell load is 125 kbps/cell.

To summarize, the soft handover gain and load factor will have an effect on both coverage and capacity. The soft handover gain in turn will assist in improving the link budget and in addition a high load factor will decrease the coverage area, increase the interference levels and at the same time limit capacity.

2.3.3 Cell Count Estimation Model

Figure 2.3 illustrates a typical cell count estimation model. As mentioned previously, an approximation of the required number of cells can

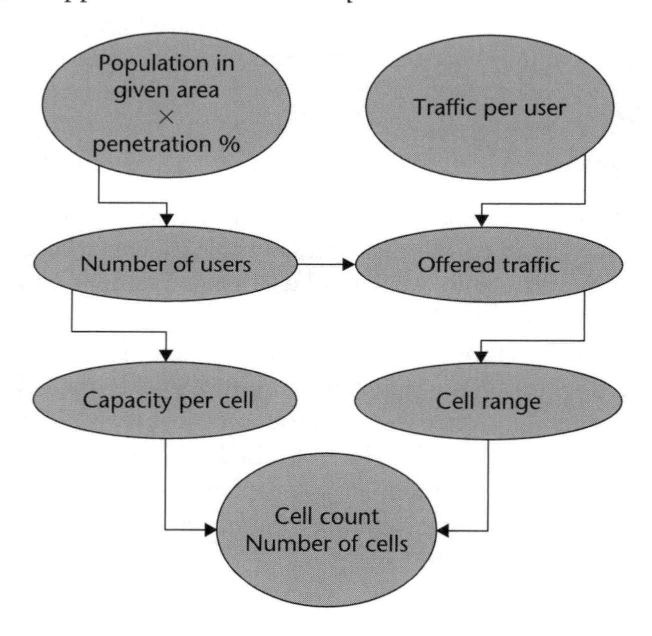

Figure 2.3 Cell count estimation model.

be calculated by multiplying the user penetration by the population. Once the number of users and offered traffic per user is known, we can calculate the total offered traffic. When range and cell capacity can be established, an approximate number of cells can be determined.

2.3.4 Area Coverage Probability

Area coverage probability refers to the probability of users accessing the network at the periphery of a cell with respect to a pre-defined outage criterion. The types of services and volume of data rate usage will also have an effect on the coverage probability.

2.3.4.1 Area Coverage Probability – Outage

A reduced probability of signal outage will require smaller cells, and therefore increased costs. So a reasonable outage probability target is considered to be around 5 to 10 per cent, which will coincide with a 90 to 95 per cent availability/coverage probability. Since there are likely to be numerous and various types of services available, this will affect the coverage probability.

Area coverage probability is also related to outage. The probability that the radio network is unable to reach the required QoS target is defined as outage. Outage is a drop in the required QoS below a pre-defined target. An outage can be characterized as the probability that shadowing and pathloss surpass the difference between the required received signal level and the maximum transmitted power. The illustration in Figure 2.4 shows a standard plot of a log-normal distribution with a standard deviation of 8 dB. The peak area of the distribution gives the outage probability. If a 10 per cent outage probability is to be achieved, or in other words a 90 per cent coverage or availability probability is required as shown in Figure 2.4, a 10.3 dB shadowing margin will then be needed. For a 5 per cent target, a 13.2 dB margin will be required, and finally an 18.6 dB margin for an outage target of 1 per cent. Therefore it is apparent that increasing the outage target leads to both reduced range and rather high margins. A shadowing plot, which typically consists of a standard deviation of 10 dB, should also be taken into account, as any increased shadowing will result in a higher probability of large shadowing values and consequently higher margins will be required. Cell periphery probability or cell area probability will be defined as the actual coverage probability.

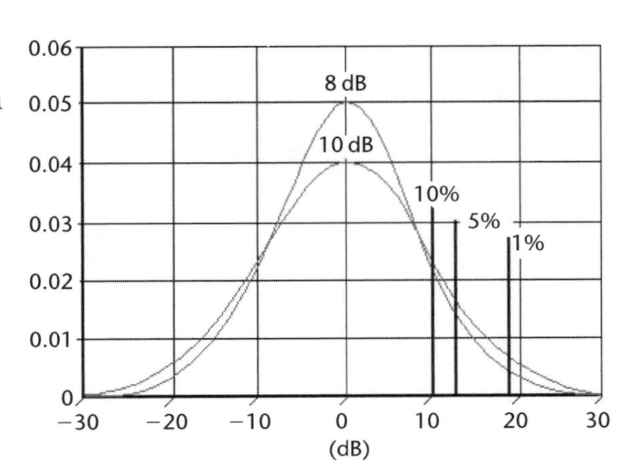

Figure 2.4 Shadowing margin and log-normal distribution.

The cell periphery probability conveys the fact of the probability that at the cell periphery, the UE will receive a signal above a certain threshold, such as the RF signal strength. The threshold value must be any value that provides an acceptable signal under fast-fading conditions, and does not necessarily have to be the receiver noise threshold.

2.3.4.2 Cell Edge Coverage Probability

Cell edge coverage probability is best thought of as the probability that a UE will receive a signal from a base station antenna, above a specified threshold value at the cell periphery. The threshold value will need to be able to provide a reasonable signal under fading conditions, therefore it does not necessarily need to meet the receiver noise threshold. The relation between cell periphery probability and area is illustrated in Figure 2.5, where 'y' is the pathloss exponent and σ (sigma) is the standard deviation of the shadowing given in decibels. The pathloss exponent is defined as the rate of increase of the pathloss. It can also be thought of as a symbol showing what power a quantity is raised to, hence it is not defined here as a specific unit of measure. A typical value for the pathloss exponent in a macrocell environment is deemed to be 3.6, and for standard deviation as just seen, will be 8 dB. Therefore, σ/y is 2.22 and taking a boundary coverage probability of 75 per cent this will equate to area coverage of 90 per cent.

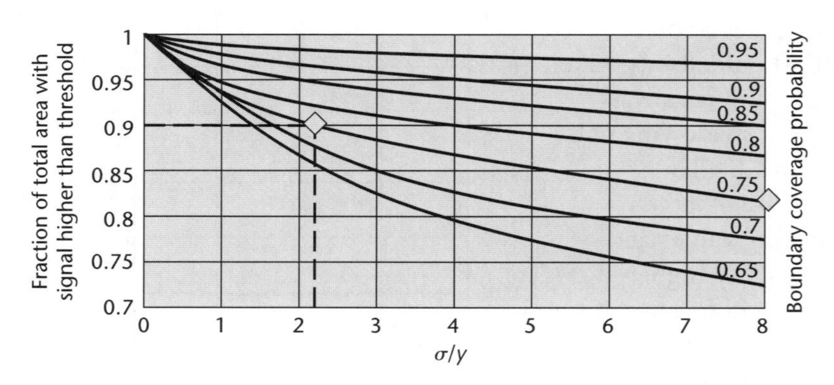

Figure 2.5 Curves relating the fraction of total area with signal higher than threshold as a function of probability of signal higher than threshold on the periphery of the cell. Original source: W-CDMA for 3G Mobile Communications, *Ojanpera and Prasad, Artech House Pubs.*

2.3.5 Blocking

Blocking is the probability of a blocked call on the first attempt and is not service specific. Soft blocking occurs when a cell has run out of capacity and users are not able to gain access to the network. A good example of hard blocking is a lack of fixed network resources, such as transmission lines, a lack of channel elements (covered later), or possibly a lack of available switching capacity. A channel element can be thought of as 'one' user in 'one' snapshot in time transferring '*x*' kbps of data. A general rule when attempting to dimension a cellular system would be to assume a blocking probability of 2 per cent. A lack of channel elements or a lack of fixed network resources, such as transmission lines or radio network controller (RNC) capacity, could cause a first call attempt to be blocked. So a shortage of system resources will cause blocking to occur. Hence each network element must be dimensioned according to the traffic demands and a blocking probability of 2 per cent is to be assumed.

2.3.5.1 The Poisson Distribution

The Poisson distribution is used to evaluate the probability of all channels being busy. The Poisson distribution is compared with the Erlang-B value for the same number of circuits and grade of service. The main Erlang traffic models consist of the Erlang-B model, the extended Erlang-B model and the Erlang-C model. The Erlang-B model is the most common model in use and can describe the required traffic circuits needed during the busy hour. This model assumes that all blocked calls are discarded. The 'extended' Erlang-B model is similar to the Erlang-B model, but the assumption here is that a percentage of blocked calls, or calls that cannot be connected due to the network being busy, are included, and the retry percentage of blocked calls can be specified here. Finally, the Erlang-C model assumes that all blocked calls remain within the system until there are enough available resources to handle them; basically the blocked calls are queued.

The fundamental difference between the Erlang-B and the Poisson equations is that the Poisson distribution assumes that blocked calls are put into a queue instead of being discarded. So if the blocking probability and the traffic density (the offered traffic) have been evaluated, then it is possible to calculate the required amount of network resources.

Trunking efficiency is also known as channel utilization efficiency and can be calculated as follows:

$$\text{Efficiency (\%)} = \frac{\text{Traffic in Erlang}}{\text{Amount of channels}} \times 100\% \tag{2.3}$$

So if our blocking probability is known, this can be used to calculate the traffic in Erlangs for the number of channel elements required, as described above. This will then give us our trunking efficiency. The illustration in Figure 2.6 shows a trunking efficiency plot for a standard assumed blocking probability of 2 per cent. This graph illustrates the relationship between the efficiency of the channel utilization and the number of channels.

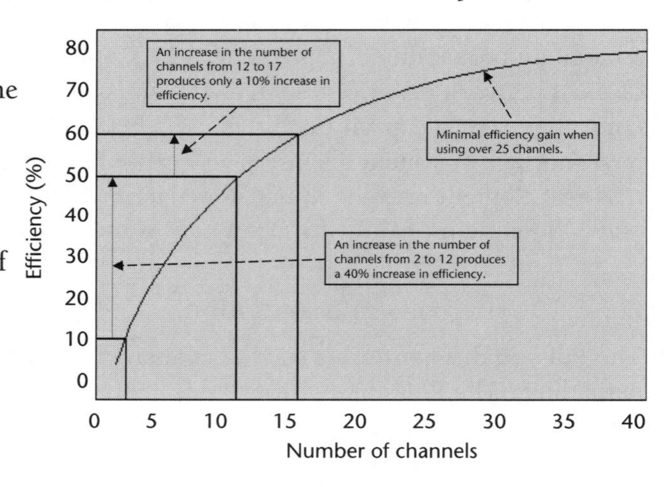

Figure 2.6 Trunking efficiency (channel utilization efficiency).

2.3.6 Simulation Analysis of Network Design Plan

Cell capacity can be estimated with a W-CDMA simulator. It is best to use separate link and system-level simulators to decrease the complexity. One example is the communications of simulation and system analysis program (COSSAP) manufactured by Synopsys Incorporated. COSSAP is a system design tool, which includes a suite of end-to-end configurations of reusable algorithmic models, which are required for the design and verification of W-CDMA systems. Within these models are complex algorithmic building blocks, with an extensive set of configurations representing different aspects of the W-CDMA system. Hence, this tool has the ability to enable the algorithmic verification of the link level for W-CDMA transmission systems under the current definition of the standards. It is known as a stream-driven simulation program, as it is based on data flow models, which support fast stream-driven simulation and co-simulation for verification. The co-simulation occurs through the co-operation of the stream-driven simulation instruction set simulators (ISS) for both the software and hardware. This kind of co-simulation verifies different pieces of a system's implementation, primarily from a functional point of view, whereby the software and hardware architecture is not fully represented in detail. With COSSAP, no timing control between simulation elements is required and it is therefore able to support asynchronous operation. The four critical factors that determine the supported capacity are as follows:

a) Traffic characteristics (e.g. burstiness and variability);
b) User data rate;
c) QoS requirements;
d) Outage probability.

A higher data rate will result in fewer users being able to use the network. So the better the network quality, the smaller the interference outage probability. However, better quality radio connections will eventually consume more resources, which means fewer subscribers will be able to use the system.

A block diagram of a system-level simulator is illustrated in Figure 2.7 which can be broken up into four main areas: creation of simulation settings, creation of gain matrix, handover, and power control. As can be seen in the aforementioned diagram these four blocks make up the basis of the system simulator model. This consists of feeding the output of the simulation settings block into the gain matrix block, which in turn feeds into the power control block. The results are then fed back into a continual loop and into the original simulation settings block. The entire process is repeated in an iterative fashion to ensure credible results can be achieved. Once the chosen simulation program has been executed, it should provide outage probability results, thus indicating the number of users that possess a worse than average signal-to-interference ratio than is actually required to be able to support the required traffic.

In summary, this section should provide the planner with the required knowledge to initiate and develop the network plan taking into account the macrocell/microcell environments with respect to both capacity and

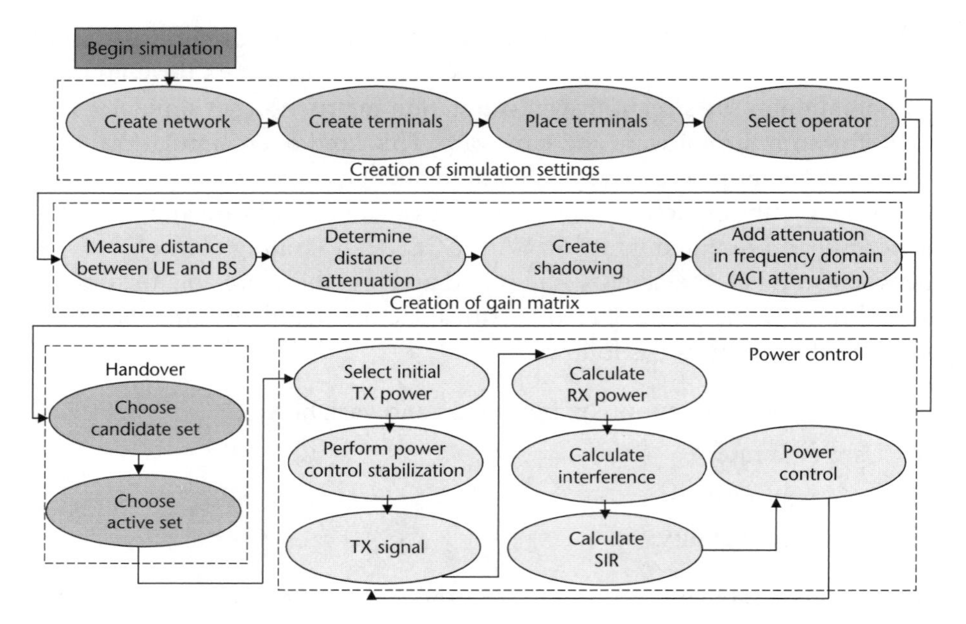

Figure 2.7 System-level simulator flowchart.

coverage, the high data rate coverage that will be required for microcell and indoor environments, the co-location with a 2G system, and finally migratory traffic. All these key issues will need to be carefully planned to ensure a stable and efficient network can be achieved. Traffic analysis, estimating cell coverage and cell count have also been discussed, along with how to perform simulation analysis.

Once these issues have been resolved and the network has been constructed according to the plan, the next step to attaining a fully functional network will be optimization, discussed in the next section.

2.4 Optimization

Network optimization involves iteratively collecting data that will provide an accurate picture of how the network is performing. Optimization will require drive-testing tools that have a high functional capacity. A few examples of the higher functional requirements needed consist of analysis of the primary synchronization and secondary synchronization code, power measurements, scrambling code measurements, delta power measurements, and timing/drift and overlap measurements. Both voice and data tests will need to be performed simultaneously on both the co-existing 2G network and the newly implemented 3G network. The data consists of the forward and reverse link conditions (signal strength), bit or frame error rates, ACI, dropped calls, failed call initialization, and the global positioning satellite (GPS) co-ordinates. It is critical to obtain this data to see a complete picture of how the network is performing.

Summary of Part I

This concludes the first part of the book which has covered the basis of 3G network planning and has introduced the all important CCQ model to the reader. As these three critical parameters interact with each other, the basis and heart of 3G network planning all relate back to achieving the optimal balance with respect to CCQ. In addition W-CDMA network planning has been covered showing how to perform a logical process of developing the procedures required and executing them to ultimately attain a workable plan. Part II deals with how to plan with regard to the first main parameter, capacity.

PART TWO

Capacity and Network Planning

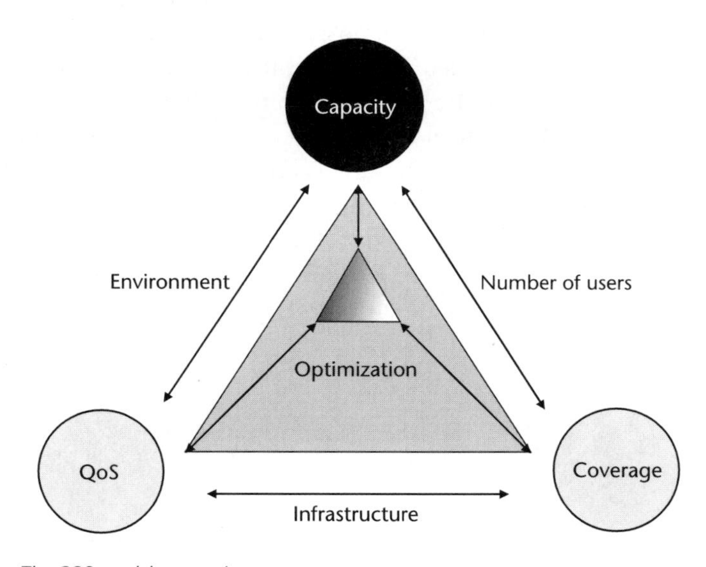

The CCQ model – capacity.

Introduction

This section covers one of the key parameters within the coverage, capacity, and quality-of-service (CCQ) model, namely capacity.

As the relationship between coverage and capacity complicates the issue with regard to both network planning and capacity estimation, simulations must be performed to provide credible results. With regard to capacity this in turn involves taking into account both the dimensioning and transmission requirements. These subjects are covered throughout this section, along with detailed network planning explaining soft capacity, capacity enhancements, soft handover planning code requirements, and their effect on capacity. In addition also discussed are network dimensioning, channel element planning, transmission planning, and the associated transmission mediums that will be utilized.

As the major characteristics of a universal mobile telephony system (UMTS) network will comprise high volume traffic, high speed data, and inevitably in the long run lower pricing plans for the subscribers, it is imperative that these networks must meet the requirements to enable adequate capacity within the macro-, micro-, and picocells which will create the network. To ensure this is possible the planner must be able to provide an accurate and detailed analysis of the proposed network, thus ensuring a stable system can be provided from the launch. With these factors in mind, it is also critical that adequate coverage can also be achieved which is dealt with in the next section of this book, Part III, entitled Coverage and Network Planning.

CHAPTER THREE

Detailed Network Planning

This chapter explains the detailed issues associated with network planning, and covers control channel power planning, radio frequency (RF) characterization, soft handover (parameter) planning, and inter-frequency handover planning. In addition, it discusses how detailed network planning needs to consider iterative coverage evaluation, pilot power planning, how hierarchical cell structuring will be implemented and the issues with code planning.

Detailed network planning requires a wealth of information as previously discussed ranging from the coverage, capacity, and quality-of-service (QoS), abbreviated as CCQ, requirements to interference levels, transmit power, multiple bit rate services, and the different delay throughputs. Statistical analysis does not provide such accurate results for calculating gains and margins because both the coverage and capacity are too interrelated. Simulations based on an iterative approach provide the most accurate picture by creating 'snapshots' in time. This is done by performing both statistical distributions and numerous iterations, which accounts for the randomness of RF propagation. This is the basis of Monte Carlo simulations which are able to produce statistical results of how the network would perform. These latter aspects are covered in more detail later in Chapter 10.

Detailed network planning initiates with the cell count, followed by the RF propagation planning for each of the cells. This will be performed with a software RF propagation prediction tool selected by the planner. There are various tools available in the market, such as Planet, Odyssey, Atoll, etc. The RF environment for each sector and cell is highly variable and dependent on many factors, such as the base station power output, height, location, antenna tilt, and the clutter present. 'Clutter' refers to obstructions that affect how the RF propagates in a cell. Examples of clutter are buildings, woodlands, forests, and uneven terrain. Just as with second-generation (2G) networks, construction restrictions, costs, and local regulations make it impossible to build an ideal third-generation (3G) network. Effective network planning means finding the optimum compromise between an ideal configuration and the unavoidable constraints that society and the environment place on the design. These aspects are all covered in this chapter and Part V – Optimization and Network Planning.

3.1 Control Channel Power Planning and Its Effects on Coverage and Capacity

Within a wideband-code division multiple access (W-CDMA) system the transmitted power levels of the control channels will have an effect on the capacity of the network. Control channels determine the coverage or reach of a sector. A user equipment (UE) is unable to access a network if it cannot utilize the control channels. The total transmitted power depends on the relative allocations of power to the pilot, paging, synchronization, broadcast, and traffic channels. The base stations are synchronized in the cdma2000 systems (refer to Chapter 1), so considerably less power is allocated to the control channels relative to the traffic channels and thus it is more efficient in moving payload data than W-CDMA extension of global system for mobile communications (GSM). The 1x radio transmission technology (1x-RTT) cdma2000 system is also a frequency division duplex (FDD) mode system. Its increased efficiency is realized by the synchronized base stations and the resulting diminished need for bandwidth allocated to overhead and control. The selection of CDMA radio technologies is governed globally by the 3G partnership projects 1 and 2, respectively 3GPP1 and 3GPP2, as discussed in Chapter 1. The 3GPP is the governing body established to ensure global control of the 3G specifications defining the network functionality, procedures, and service aspects. However, it was decided to split this organization into two bodies, 3GPP1 and 3GPP2. This was initiated mainly to allow the 3GPP2 to focus on the multi-carrier (MC) cdma2000 and cdmaOne™ technologies, leaving the 3GPP1 clear to deal with the W-CDMA universal mobile telephony system (UMTS) variant. Europe and most of the Far-East have selected the W-CDMA technology, whereas the USA, Korea, and a few other countries have opted for the cdmaOne™, and/or the cdma2000 variants (cdmaOne™ is a trade mark of the CDMA Development Group (CDG)).

With W-CDMA significant numbers of overhead channels are required to account for the lack of synchronization among base stations. When using W-CDMA the total transmitted power (in watts) depends on the relative allocations of power to the pilot, paging, synchronization, broadcast, and traffic channels; therefore overall coverage is related to power allocation. Since the UE must be able to decode neighbouring base stations before it can enter a soft or a softer handover state, control channel coverage must be greater than traffic channel coverage. The synchronization channel will consume approximately 5 per cent of total base station power. The pilot power will be set by the planner and will be dependent on the cell coverage area required (covered later in Section 3.6 of this chapter). The broadcast channel carrying the cell information must be processed before the UE enters the coverage area of a cell. If an 8 kbps speech traffic channel were in operation, then around 3 dB more power would be required for the control channels to maintain the connection.

3.2 Capacity and Soft Handover Parameter Planning

Soft capacity is the maximum capacity of the active connections limited by the amount of interference present in the air interface. As the interference levels will constantly be variable, there cannot be an actual fixed value for the maximum capacity. The average number of channels per cell is smaller than the total channel pool since the neighbouring cells share part of the same interference and thus more users can be connected to the cell with the same blocking probability. An indication of soft capacity can be seen with the use of the illustration in Figure 3.1, which depicts that a higher cell capacity can be achieved by one cell, assuming there is a lower level of interference present within the adjacent cells. With a small number of channels per cell, such as for real-time users demanding high data rates, a low blocking probability can be guaranteed if the average loading is relatively low. Therefore, when few users demand high data rate services,

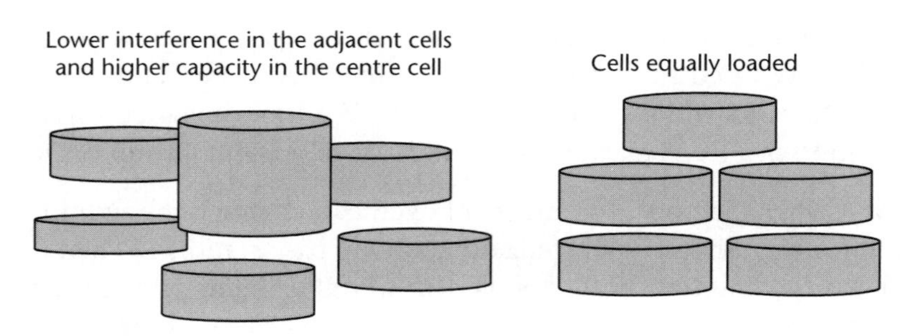

Figure 3.1 Interference sharing between cells (based on WCDMA for UMTS, *Holma and Toskola, Wiley).*

relatively few channels will be allocated per cell and a low blocking probability can be guaranteed when the average loading per cell is kept low. With the average loading held low, extra capacity is present in adjacent cells. A cell can borrow this excess capacity; doing so shares interference among the cells and creates additional soft capacity.

3.2.1 Soft Capacity Calculations

With regard to soft capacity calculations, it is assumed that the number of users are equal in all cells, but that the connections start and terminate at different times. Regarding a connection arrival interval, the Poisson distribution must be considered, as this can be used when calculating Erlang capacities. Poisson distribution will enable an evaluation to be made of the probability of all channels being busy. The fundamental difference between the Erlang-B and the Poisson equation is that in the Poisson distribution we assume that all blocked calls are put into a queue, instead of being discarded. Therefore, there is extra soft capacity available if there are fewer subscribers in the adjacent cells. We can estimate the uplink soft capacity from the total interference at the base station's receiver. This total interference includes both the current cell and the adjacent cell interference. Thus, as shown in the calculation below, the total channel pool can be achieved by taking the number of channels per cell in the equally loaded scenario and multiplying them by $1 + j$, which will then result in the single isolated cell capacity. For the purpose of this calculation 'j' can really be thought of as the adjacent cell to current cell ratio of interference, and is a function of cell isolation. The larger channel pool, which equates to the interference pool, is then calculated with the standard Erlang-B formula. The Erlang capacity acquired is then equally shared

Table 3.1 Assumptions for soft capacity calculations.

Bit rates	Speech 12.2 kbps Real-time data 16–144 kbps
Voice activity	Speech 67% Data 100%
Eb/No	Speech: 4 dB Data 16–32 kbps: 3 dB Data 64 kbps: 2 dB Data 144 kbps: 1.5 dB
Blocking probability	2%
Noise rise	3 dB (=50% load factor)
j	0.55

Source: WCDMA for UMTS, Holma and Toskola.

between the cells. To estimate the soft capacity, the following procedure (also note the assumptions noted in Table 3.1) must be followed:

a) Calculate the number of channels per cell, '*x*' in the equally loaded scenario; this will be based on the uplink load factor equation (uplink load factor equations are covered later in Chapter 6).
b) Multiply the number of channels by $1 + j$, this will give the total channel pool in the soft blocking situation.
c) From the Erlang-B formula, calculate the maximum offered traffic.
d) Divide the Erlang capacity by $1 + j$.

Soft capacity calculations

$$\text{Soft Capacity} = \frac{\text{Erlang Capacity with Soft Blocking}}{\text{Erlang Capacity with Hard Blocking}} - 1$$

$$\begin{aligned} \mathrm{j} + 1 &= \frac{\text{Adjacent Cell interference}}{\text{Current Cell interference}} + 1 \\ &= \frac{\text{Adjacent Cell interference} + \text{Current Cell interference}}{\text{Current Cell Interference}} \\ &= \frac{\text{Isolated Cell Capacity}}{\text{Multicell Capacity}} \end{aligned} \tag{3.1}$$

3.2.2 Soft Capacity Assumptions

Some typical assumptions required for performing soft capacity calculations are shown in Table 3.1. The soft capacities obtained with

Table 3.2 Soft capacity calculations.

Data rate (kbps)	Channels per cell	Hard blocked capacity (Erl)	Trunking efficiency (%)	Soft blocked capacity (Erl)	Soft capacity (%)
12.2	60.5	50.8	84	53.5	5
16.0	39.0	30.1	77	32.3	7
32.0	19.7	12.9	65	14.4	12
64.0	12.5	7.0	56	8.2	17
144.0	6.4	2.5	39	3.2	28

Source: WCDMA for UMTS, Holma and Toskola.

respect to the Erlang/cell based on an uplink loading equation are shown in Table 3.2 (uplink loading equations are covered in Chapter 6). The trunking efficiency, which is sometimes known as the channel utilization efficiency, is defined as the hard blocked capacity divided by the number of channels. If the capacity is hard blocked it is limited by the amount of hardware, and the Erlang capacity can be obtained from the standard Erlang-B model. If the trunking efficiency can be maintained at a low level, then the average loading will remain low; this will be beneficial as more capacity is then available which can be borrowed from the adjacent cells.

3.2.3 Soft Capacity as a Function of Bit Rate

Figure 3.2 illustrates that more soft capacity is available for higher data rates than for lower data rates. As can be seen in the illustration there will be around 22 per cent of soft capacity available when users within the cell are employing a data rate throughput of 144 kbps. Soft capacity in general will depend on the RF propagation in the coverage area and therefore will be dependent on the quality of the network planning. This will have an effect on the value of the current cell to adjacent cell ratio, or in other words '*j*'. With higher bit rates a lower trunking efficiency exists and therefore there are fewer 'channels per cell'; hence the average loading is less and more capacity can be borrowed from the adjacent cells, so allowing more soft capacity to be available.

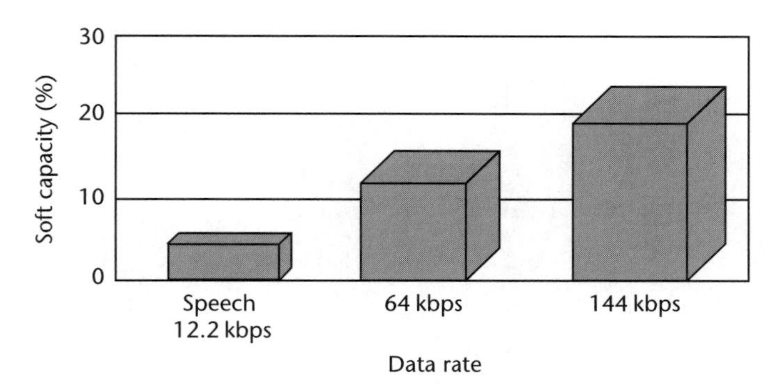

Figure 3.2 Indication of soft capacity for real-time connections.

3.2.4 Soft Capacity and Radio Resource Management

Radio resource management (RRM) will ensure that a high capacity can be achieved within one cell when the adjacent cells are experiencing a lower loading. For this to be possible, it is imperative that the RRM algorithms are based on wideband interference and not based on the amount of data throughput or the number of users holding an active connection (this is covered in more detail in Chapter 8). If the interference based RRM algorithms are utilized, then soft capacity will also be available in the downlink.

3.3 Capacity Enhancements with Inter-frequency Handover Planning

Coverage areas that are constructed in a hierarchical structure consist of microcells overlaid by macrocells. This can also include hot-spot cells with more carriers than the neighbouring cells. In this case inter-frequency handovers will be required to maintain these connections. With the hot-spot scenario illustrated in Figure 3.3, the intracell handover within the hot-spot cell, once performed, would be classed as an inter-frequency or hard handover. So initially an inter-frequency handover (hard handover) will be executed from frequency 1 to frequency 2. The execution of inter-frequency handovers can be achieved with the utilization of two systems, which are described later in Section 5.8 and are known as the compressed mode (slotted mode) and the dual receiver. The dual receiver has a

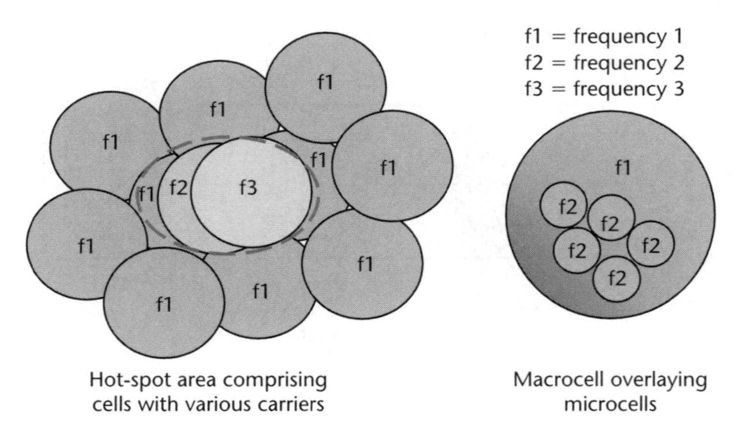

Figure 3.3 Inter-frequency handover scenarios.

somewhat simple procedure, but is rather costly to implement as it includes a diversity receiver, or two receiver chains. One of the receivers performs measurements on the alternative carriers available to the terminal. The alternative system to this is the compressed or slotted mode which comprises a relatively simple receiver and is therefore more cost efficient. However, one of the disadvantages with the compressed mode system is that the receiver algorithm is unable to compensate effectively for the burstiness of the required signals, and this can cause performance degradation.

3.3.1 Microcells and Macrocells on Different Frequencies

It is more straightforward to manage overlaying and underlaying micro- and macrocells which operate on different frequencies, because the interference is minimal and consists mainly of adjacent channel spillover. The downside here is that each cell layer requires its own frequency, so the spectrum requirement is quite large.

Adjacent channel attenuation refers to the amount of attenuation that the *interfered-with* user experiences, if it is received at the adjacent channel. The increased adjacent channel interference experienced governs the degradation of the spectrum efficiency. This is caused due to the non-linearity of the power amplifiers. Table 3.3 shows an example of the

Table 3.3 Sample ACI masks.

	ACI attenuation (dB)	Centre frequency spacings (MHz)
Micro- to macrocells	33	5
	35	6
	43	7
	52	8
	58	9
	62	10
Macro- to microcells	27	5
	29	6
	32	7
	36	8
	45	9
	53	10

For example, ACI attenuation at 5 MHz = 6 dB (33 − 27 dB) and 8 MHz = 16 dB (52 − 36 dB).

required adjacent channel interference (ACI) attenuation from which the following examples can be seen:

a) At a centre frequency spacing of 5 MHz, this equates to 6 dB (33 − 27 dB).
b) At a centre frequency spacing of 8 MHz this equates to 16 dB (52 − 36 dB).

3.3.2 Microcells and Macrocells on the Same Frequency

It is possible to have both the micro- and macrocells operating on the same frequency. In this case interference from any adjacent cell is likely to occur due to the processing gain. The processing gain is the ratio of transmitted bandwidth to the information bandwidth. Intra-frequency interference is influenced by power control, covered further in Chapter 5. Spatial isolation, which can be thought of as frequency re-use within multiple coverage areas that are physically isolated from each other, will also have an effect on the intra-layer interference. As can be seen in Figure 3.4, the location of the crossover point will vary due to the actual receiver and transmitter antenna heights. As the microcell base station antennas are generally installed lower to the ground, the coverage area will be somewhat reduced. Therefore, the crossover point for the microcell will

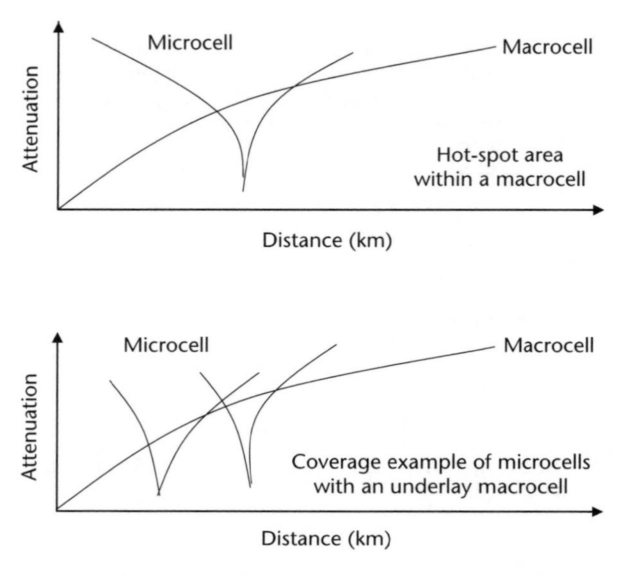

Figure 3.4 Spatial separation of umbrella cells and underlay microcells. Source: WCDMA for 3G Mobile Communications, *Ojanpera and Prasad.*

occur relatively close to the base station. Hence, as the macrocell antennas are positioned at a higher level, the signal from a macrocell base station will attenuate at a slower rate than the signal from a microcell base station. A major dilemma that is likely to occur can be attributed to the delay in the overhead signal processing. However, if the UE arrives in the microcell area, it is most likely to be holding simultaneous connections to both macro- and microcell base stations. This can be caused by inaccuracies in the handover measurements and the time taken during the signal filtering occurring within the handover measurement process, explained in more detail in Chapter 8.

Careful soft handover parameter planning should be exercised, as the size of the handover region must be large enough to prevent any hysteresis occurring. Hysteresis is sometimes described as 'ping-ponging', which refers to the continual handoffs repeatedly occurring back and forth between sectors or cells. This is detrimental to the system performance because the signalling load is increased and valuable resources are unnecessarily wasted. A standard default value of 4 dB relative to the 'add threshold' and the 'drop threshold' can be applied regardless of the vendor or technology platform and this will prevent 'ping-ponging' from Cell 'A' to Cell 'B', unless Cell 'B' is detecting the UE better than Cell 'A' by 4 dB. These concepts are covered further in Chapter 7. This problem is particularly applicable if the UEs are moving fast as this further complicates the UE's measurement process. Furthermore, fast fading must also be taken into account, as this will have an effect on the closed loop power control for the UE. In addition, where a considerable amount of prospective candidate base stations are present, there will be a noticeable delay before the pilot signal of a prospective handoff base station can be measured. The signalling required between the UE and the base station can only be initiated after the handover decision has been made. So a delay is likely to occur before the UE and the radio network controller (RNC) acquire and process all the handover signalling requirements.

Delays of a few seconds can easily occur due to filtering and pilot measurements. Filtering will be applied with respect to the pilot energy-per-chip/spectral density (Ec/Io). This will consist of an average mean being computed from a certain number of the previous measurements taken before the pilot Ec/Io is used by the active set update algorithm in the UE. To give an example, suppose a UE is travelling in a vehicle at a speed of around 60 km/h. It will therefore be moving at a rate of approximately 17 metres/second. If the UE is connected to a microcell with a coverage radius of 150 meters, then only a 9 seconds delay on handover is required before the UE reaches the centre of the cell. Fast moving UEs in overlay/underlay cells will not be subject to these hysteresis and filtering problems, as any overlay/underlay cell will be operating on a different frequency, therefore the handover will be a 'hard' handover as experienced in GSM.

3.4 Changes in Capacity and Quality of Service with Non-uniform Traffic

System performance is generally reduced when traffic is of a non-uniform nature. Increased interference causes a capacity reduction within the cell. This can be attributed to the distribution of the various users who are transferring different data rates throughout the network. A balance between the use of both uniform and non-uniform traffic would be ideal, as the increase in interference diminishes the quality of the connections, particularly in high volume traffic areas. Alternatively, should the volume of traffic diminish, then the QoS available to the remaining users will increase due to the reduction in interference levels.

3.5 Coverage Changes with Adaptive Control of Cell Radius

Antenna directivity, controlling the cell radii, and the uplink received power threshold can all be used to assist in enhancing the system efficiency. Accurate planning is required if antenna directivity is to prove beneficial, as the interaction with handovers must be considered and the proposed coverage area for each sector must be correctly targeted. The cell radius can be increased if the signal-to-interference ratio (SIR) is greater than actually required. Alternatively, if a lower than required SIR is experienced, then the opposite applies and the cell size shrinks. The cell radius can be controlled by varying the effective radiated power (ERP) of the pilot channel. To ensure the correct balance of the cell radii in both the uplink and downlink, the received power threshold in the uplink can be adjusted.

3.6 Pilot Power and Pilot Pollution

Pilot pollution occurs when no pilot is strong enough to let a UE establish a connection, that is where there is no dominant server present. This situation occurs when there is a good UE received signal, but a poor Ec/Io interference ratio and a poor forward bit error rate (BER) of about 3 per cent or more. To ensure pilot pollution can be efficiently managed, a cell plan must be generated in which only one dominant pilot is present. There are many methods that can control pilot power. Some of these are down-tilting antennas, increasing the coverage of sectors/cells, and adjusting the ERP of the pilot channel. The actual pilot strength or Ec/Io can be used to identify problems. The UE requires a sufficient Ec/Io to ‘lock on’ to, or remain active in the system. A UE may not even be able to initiate a network connection in an area with an excessively low Ec/Io which is typically caused by low pilot ERP or excessive pathloss. One method that could be used to provide a possible solution for a low Ec/Io would be to increase the pilot ERP. Pilot pollution contributes to forward link interference. Forward overhead transmissions are sometimes described as pilot pollution, as the coverage of the control channels (e.g. the synchronization, broadcast, and pilot channel) must be greater when compared to the traffic channels. The pilot powers are generally speaking the highest among all overhead channels by up to 30 dBm. Since the 'Io' (noise spectral density from all the interferers) is often the same as the total overhead power from all the neighbouring base stations, reducing their collective power reduces this kind of interference. Other factors causing forward link interference to be considered are interference power due to forward traffic channel transmission of the home base station, and interference power due to forward traffic channel transmissions of other base stations and external RF interference.

All other signalling and control channels will typically be transmitted at a lower fixed offset from the pilot. This will vary and be dependent on many factors, such as the data rate in use, interference levels, volume of traffic in the sector/cell, and the UE's location. The pilot can be thought of as a beacon, which aside from its radio technique utilities such as a phase reference, is responsible for the following:

a) Adding cells to the active set in a handover scenario.
b) Providing a channel estimate at the UE's receiver.
c) Setting the maximum coverage of the cell.

Reducing the pilot strength in an isolated cell (e.g. a cell on the periphery of the UMTS coverage area) will cause a reduction in that cell's coverage area. As shown in the illustration in Figure 3.5, if there is more than one cell providing contiguous coverage and the pilot power is reduced, this will create the effect of moving the periphery of the cell towards the other cell. However, if the pilot power is increased, this will have the opposite effect of reducing the periphery of the cell and therefore has a marked effect on coverage.

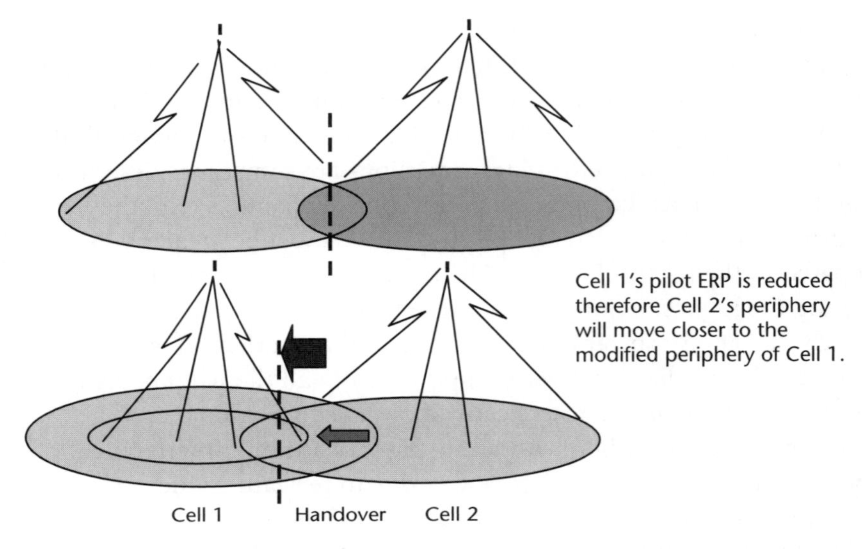

Figure 3.5 Pilot power modification.

3.7 Soft Handover Planning in Terms of Capacity and Coverage

The performance of soft handovers will have a distinct effect on the actual fade margin against shadowing and the number of users within a soft handover scenario. The fade margin is a critical factor within the link budget calculation and consequently will have a dramatic influence on the number of base stations that have to be installed. Users in soft handover scenarios will need extra channel resources, and the planner must ensure the current transmission backhaul is able to support the extra capacity. The transmission backhaul is the method of routing which is used to transport the traffic between the core network and the base station. The ultimate aim here is to minimize the fade margin and reduce the number of users in a soft handover state, while simultaneously ensuring an acceptable level of QoS. The actual number of users will depend on the types of services used (e.g. high data rate services such as video-to-video and lower data rate services such as e-mail delivery). Handover thresholds must be taken into account and can be set by the operator. These consist of 'handover add' and 'handover drop' thresholds. Once the signal strength of the second base station exceeds the 'add' threshold, the UE is then in a soft handover state. When the original base station's signal strength falls below the 'drop' threshold, the first base station no longer maintains a connection to the UE and is dropped from the active set. The UE must negotiate an addition to the active set and subsequently there must also be adequate backhaul capacity available. A handover threshold is explained in more detail in Chapter 8.

Antenna orientation and antenna tilts should be taken into account when performing the handover parameter planning, as this will influence the number of users in a soft handoff scenario and consequently will affect the coverage and capacity of the area. To ensure the most favourable active set size is achieved, particular attention must be paid to the handover thresholds, the number of available RAKE fingers and the RF environment. A rough estimate would suggest that including more than three base stations in an active set would not improve the soft handover gain due to the increased signalling load generated. In certain deployment scenarios an active set may be able to support up to six connections. This will be beneficial in areas experiencing high volume data traffic, as it will assist in spreading the load throughout the sectors/cells available, thus causing a reduction in interference levels and, in turn, an increase in capacity. This may occur in large sectored sites, (e.g. six sectors). In addition, it is possible to increase the number of 'fingers' in a RAKE receiver where more 'fingers' are added, and this will assist in improving the 'reconstructed' signal quality. Adding more 'fingers' will most likely result in an increase in the physical size of the RAKE receiver, as well as a definite increase in the power requirement. As each 'finger' can be considered as a separate receiver, the quantity of the receivers will increase. This increase in the amount of hardware will have a detrimental effect on spatial requirements and cost of equipment. Apart from the cost, this is not considered a significant problem for the base station, as the size and power constraints are not such an issue, but these limitations will be a problem at the UE end.

3.8 Hierarchical Cell Structures

A hierarchical cell structure (HCS) is a system where two different cell types can co-exist on top of each other as shown in Figure 3.6. To achieve both a high capacity and high spectrum efficiency, small cells are required.

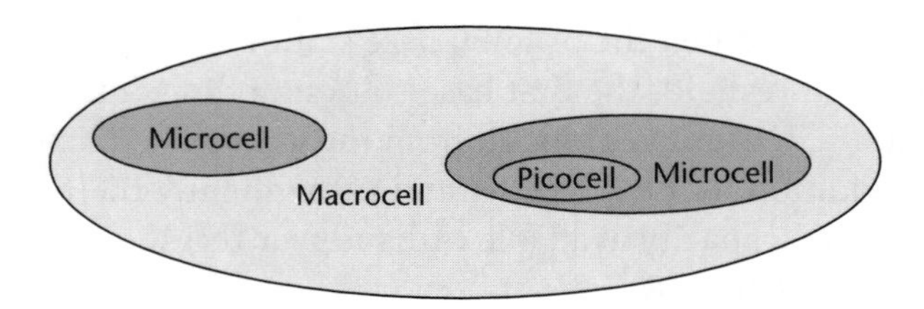

Figure 3.6 Hierarchical cell structures.

It should be noted that it is beneficial to 'steer' traffic towards smaller cells as this will improve the spectral efficiency. A higher noise rise is also allowed in microcells as there will be much less of a coverage reduction in small cells as compared to macrocells. In addition, small cells are well suited for high capacity UEs with low mobility, whereas high range cells are well suited for low capacity and high mobility.

There are two methods for HCS design within W-CDMA. The first approach consists of co-existing hierarchy layers in the same frequency band. Where different cell area layers are employed, the users can be differentiated by signal fading, handovers, and possibly code resources. The second and more attractive method consists of all hierarchical cell layers operating on different frequencies and therefore soft handovers will not be possible. A hard handover is similar to a GSM-type handover, where the connection is momentarily cut and changed to another frequency (for more information on GSM hard handovers refer to '*GSM Networks*' by Heine, Artech House (2001)). This is also relevant in a multi-operator environment where an operator has to consider interference from the adjacent frequencies belonging to another operator. Correct frequency planning will ensure that the use of hard handovers will minimize any adjacent, co-channel, or external operator interference.

3.9 Code Requirements and Their Effects on Capacity

There are two types of codes available, both long and short. There are more than enough long codes available to supply the users in any one cell. However, short codes must be more strictly managed as they are a limited resource. High data rate traffic such as video-to-video, which would

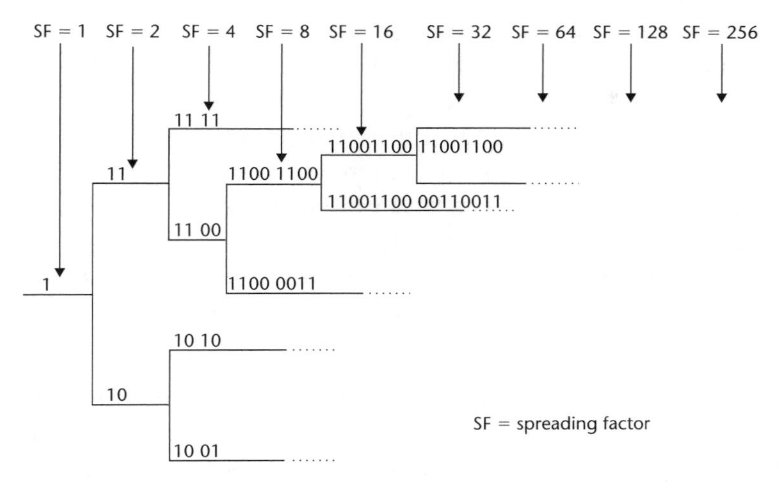

Figure 3.7 Code tree (channelization).

require the use of short codes, also requires a low spreading factor, which consequently will block a large number of the lower order codes available in the channelization code tree. A code tree is therefore required to ensure that orthogonality between the users can be maintained. An illustration of the channelization code tree is shown in Figure 3.7. There are 256 completely orthogonal channelization codes available and under certain conditions, one cell is able to make use of all these codes. However, the number of codes available is dependent on the nature of the traffic at a given point of time. In addition, the types of traffic in use may vary their data rates, hence this will cause a change in the spreading factor values, which in turn will relate to the lengths of the code(s) required. Therefore, all the 'shorter' channelization codes, which lead up to the point where the 'longer' channelization code is located within the tree, will not be available. The length of the channelization code is always the length of one symbol in chips.

So it can be stated that the greater the spreading factor, the greater the number of orthogonal codes will be available. In addition, the greater the spreading factor value, the smaller will be the base band data rate. Therefore, the user with a bearer that is using a lower data rate will consume less of the available code resources. This, together with the spectrum efficiency, will have a significant effect on the number of codes available. For the system to be able to support low data rate speech traffic, a large number of codes will be needed. To ensure channel separation is achieved, orthogonal codes are used in the downlink. In a scenario where adjacent cells are free of traffic, the bordering cell can contain a much higher volume of users and hence more codes are required. UE currently engaged in soft handover scenarios will consume codes from at least two base stations and will be dependent on the number of cells present in the active set. The same set of orthogonal codes can be allocated, but with different pilot codes for the same sector, to increase the number of codes available and assist in increasing capacity.

The following three subsections cover the types of codes in further detail: firstly, the static and dynamic codes; followed by pseudo-noise (PN) sequences; and finally, orthogonal codes are explained, along with an example of Walsh codes.

3.9.1 Static and Dynamic Codes

Static or dynamic code allocation can be used in a CDMA system. With dynamic codes, once the user is inactive, then the spreading codes are released, thus enabling the re-use of these codes by other users. If numerous users are active and transferring high data rates, then it is possible that the remaining available codes may all be consumed. Hence, the dynamic code allocation would be considered more suitable, especially where high variable data rate usage is expected. Conversely, if both the traffic and the data rate usage were to be rather low, then only a small portion of the allotted codes for each user would be required. With regard to static code allocation, if the activity factor appeared to be low, then only a minimal part of the allocated codes would be used for each connection. However, it would be possible for the base station to exhaust its code supply if similar types of users were introduced into the cell. Consequently, dynamic code allocation is more practical.

3.9.2 Pseudo-noise Sequences

Two types of codes can be considered here: orthogonal codes and PN sequences. Orthogonal codes are made from an even number of chips and the number of codes is equal to the number of chips. PN sequences contain an odd number of chips. Noise-like wideband spread signals can be generated by using a PN sequence, which is deterministically generated, but appears as a random sequence. To be of use within a direct sequence spread spectrum system, the PN codes must consist of bit sequences and have a one-chip wide auto-correlation peak, allowing synchronization and equal spreading to be achieved over the entire frequency band in use. In addition, lower cross-correlation values are required to allow more users into the system. A combination of PN sequences (e.g. Gold codes) is somewhat complex to generate. PN sequences are really direct spread sequences and are generated by linear feedback shift registers. To ensure synchronization is possible, it is recommended that the spreading code length should be a multiple of a power of two. In such a case, this would require an extension of one chip for the PN sequences. This should be performed so that the numbers of 'zeros and ones' are equal. It is possible that Gold codes may be used in the downlink for sector/cell separation. However, it would also be possible to use them in the uplink for user separation.

3.9.3 Orthogonal Codes

When considering a synchronous transmission, such as the downlink, the cross-correlation properties of orthogonal codes equates to zero. Therefore, such codes, (e.g. Walsh codes) are generally considered acceptable for channel separation. Orthogonal codes, such as Walsh codes, can be thought of as possessing good separational properties. Walsh codes consist of an even number of chips where the number of codes is equal to the number of chips. With Gold codes the number of codes is approximately the same as the code length. Code lengths of say, 127, 511, and 2047 are odd degree codes and hence a smaller value of maximum cross-correlation relative to the actual code period can be realized. User separation cannot be performed without the use of orthogonal codes, and assuming no multi-path propagation occurs, then the orthogonality will remain when the downlink signal is received by the UE, as without any multi-path propagation the downlink is synchronous. However, when multi-path propagation is present, to ensure the orthogonality remains, an overlay code can be used to reduce the interference generated by multi-path propagation. This will also result in improving the auto-correlation properties. If sufficient delay spread occurs in the radio channel, then the

UE will read part of the required downlink signal as multiple access interference. An orthogonality factor (α) of one is equivalent to perfect orthogonal use, as compared to a factor of 0.4–0.9, which is generally found in multi-path channels. These figures have been derived from experience with IS-95 systems. The orthogonality factor can be derived from the following equation:

Orthogonality factor

$$\alpha = 1 - \frac{\text{Eb}}{\text{Io}}\left(\frac{\text{Eb}}{\text{No}}\right)^{-1} \qquad (3.2)$$

where No is intercell interference, and Eb/No is expressed in absolute figures and not in decibels.

The performance of an orthogonal code is dependent on the delay spread, the pathloss and the channel profile. The performance gain will be higher if there is good isolation between the cells and a lower multi-path spread.

3.9.3.1 Walsh Codes

An example of Walsh codes is illustrated in Figure 3.8, showing how orthogonal channelization codes can be created. As can be seen in the illustration, the preceding matrix is used to build the next matrix, and the right-hand lower corner is a complement member each time, and therefore

General rule:

$$W_k = \begin{bmatrix} W_{k/2} & W_{k/2} \\ W_{k/2} & \overline{W_{k/2}} \end{bmatrix}$$

$$W_1 = (0) \Rightarrow W_2 = \begin{bmatrix} 00 \\ 01 \end{bmatrix} \Rightarrow W_4 = \begin{bmatrix} 0000 \\ 0101 \\ 0011 \\ 0110 \end{bmatrix} \Rightarrow W_8 = \begin{bmatrix} 00000000 \\ 01010101 \\ 00110011 \\ 01100110 \\ 00001111 \\ 01011010 \\ 00111100 \\ 01101001 \end{bmatrix} \text{ etc.}$$

Figure 3.8 Walsh codes.

the numerical values are inverted. Assuming the system is synchronized and no phase shifting is occurring, then these types of codes will be completely orthogonal. In theory it is possible to generate any length of code, however, the fixed length of spreading factors is the limiting factor here, (e.g. 4, 8, 16, 32, 64, etc.). With this in mind it would not, for example, be possible to use a Walsh code of 19.

The above sections describe the various codes available for UMTS planning. Each type of code possesses different characteristics that are beneficial in code planning methodologies.

3.10 Code Planning

To select the appropriate codes for use over a specific air interface, the number of codes and their correlation properties must be taken into consideration. Furthermore, correlation must be taken into account, as in an asynchronous system it is possible for codes to have a phase shift relative to each other. In this case the orthogonality is degraded and the system performance suffers. Therefore, cross-correlation properties of the PN codes will need to be optimized. Optimization of codes can be awkward as scenarios will occur where codes should be optimized both against codes of the same length and other spreading factors with different code lengths, a situation likely to have an overall effect on the performance. In the case of two UEs both using the same high data rate service, having a good selection of spreading codes available will help to reduce the likelihood of any performance degradation. In addition, a delay difference experienced by two users is likely to have an effect on the maximum correlation value, thus also causing the performance to suffer.

The planner must be aware that the synchronization requirements may have an effect on the spreading code selection available for the synchronization and access channels. This can be due to the auto-correlation properties of the codes in use, as they may affect both the time taken for channel acquisition and the performance, consequently acquisition time should be brief. Code length here is variable and will be dependent on the required user data rate and spreading factor. Spreading code selection is likely to be affected as more complex receivers are brought into use. In the uplink each user will have individual spreading code. Re-use does not become an issue with the downlink scrambling codes. As an example, if an average figure of three codes were to be allocated to each base station, then a re-use factor of 170 would be possible (512 codes/3 sectors = 170.66 possible codes). The planner should also be aware that planning for the secondary synchronization codes, which are currently under further study by 3GPP, may also be required. However, re-use in the following specific example is also not critical, as there are 64 secondary synchronization codes, thus giving a re-use factor of around 20. This example assumes that an average figure of three codes per base station (64/3 = 21.33) will provide a re-use factor of approximately 20. Any re-use factor will prove beneficial as it will enable more codes to be allocated, this in turn allowing more users onto that cell or sector.

In conclusion, the code planning will mostly be performed by the planning tool. However, the planner should be aware that the channelization codes are a limited resource and this must be taken into consideration when planning a UMTS network.

3.11 Characterization of the Radio Frequency Environment

To ensure accurate optimization the RF environment has to be taken into consideration, along with pathloss, RF propagation, intermodulation, and shadowing. The results obtained can be crucial in determining inter-operator interference, intermodulation, and background interference.

Pathloss is a critical factor and can be calculated with RF propagation models. Large-scale propagation models, which would typically be used for predicting RF propagation over large areas, determine the received signal strength by performing average power level measurements over highly separated distances. These can vary up to a few kilometres for both the receiver and transmitter at opposite ends of the path. If the receiving antenna is located closer to the transmitting antenna, or if the paths are shorter, such as tens or hundreds of metres, then a medium-scale propagation model should be used, thus ensuring any changes in the local-mean power can be resolved. Shadowing is caused by obstructions and can be thought of as the medium-scale variation of the received signal power.

In coverage and capacity limited cases, shadowing, which is discussed in Chapter 9, will have a marked effect on the network performance and hence a realistic shadowing margin must be taken into account (see Section 2.3.4.1, entitled Area coverage probability – outage).

3.12 Radio Network Testing

Once the network has been built and a 'user friendly' network deployed, it can be drive tested to ascertain its actual quality. Various high profile potential customers are often given access to the 'user friendly' network, along with both operator and vendor personnel, hence the performance and quality can be closely monitored. Useful network data is also acquired from drive tests, which consist of performing various different types of field measurements. The results are then used within the optimization process, thus enhancing network performance. There are several types of tests associated with drive testing such as signal strength measurements, coverage problem identification, customer complaint validation, co-channel and ACI testing, cell site parameter adjustments, software change testing, post cell turn-on, and post re-tune efforts. The crucial measurements are first determined by the optimization department before initiating the actual testing, and include dropped call rates, coverage limitations, pilot sets for soft handover, and quality. Therefore, the implementation of operations, testing, and maintenance facilities should be considered as a pre-requisite to running the network successfully.

In summary, the planner should now be aware of how to initiate a planning process, develop the necessary procedures and the design plan for a 3G network. Other areas covered include estimating the cell coverage and cell count, along with how to perform simulations. Soft handover parameter planning, HCSs and code requirements are also covered. This leads into the next chapter which covers network dimensioning.

Network Dimensioning

The network dimensioning phase consists of creating estimates for the required site density and the site configurations. Once initial dimensioning has been performed, it should enable the capacity requirements for the site configurations to be estimated. At this stage dimensioning encompasses the backhaul for the network or it could be described as the transmission capabilities for moving the traffic throughout the network. Strong emphasis is placed on ensuring the network, and backhaul capacity will be able to handle the predicted volume of traffic. Network dimensioning has a critical effect on the network planning exercise and, in particular, the aspect of capacity within the coverage, capacity, and quality-of-service (QoS) model, abbreviated as CCQ, as we will explore with further detail in this chapter.

Network planning parameters for universal mobile telephony system (UMTS) are somewhat complex and consist of a vast number of physical, system, and simulation parameters all of which are covered throughout this book. All these parameters relate to CCQ. Channel elements (CEs), which are covered in further detail in Section 4.1.2 of this chapter, have a fundamental effect on capacity and, in turn, the configuration and number of base stations required. The number of CEs at a specific base station will limit the maximum transmission data rate at that location. In addition, uplink and downlink load factors are discussed, along with both uplink and downlink capacity issues and, lastly, some useful capacity improvement issues. This will conclude Part II on capacity and is followed by the next section, which addresses coverage.

4.1 Dimensioning

As previously mentioned, dimensioning is used to maximize the three critical CCQ parameters. The planner will by now be aware that, dependent on the agreed design requirements, one of the three key CCQ parameters may be forced to take precedence over the remaining two parameters. For example, if the planner is forced to provide higher capacity within the proposed network, then this will affect both coverage and QoS. To ensure dimensioning is performed correctly, it cannot be emphasized too strongly that all three of the CCQ parameters must be equally considered in order to achieve the optimal solution.

4.1.1 Traffic Modelling and Services

Within a UMTS network, complexities exist with regard to the relationship between the data rates being used, and both the cell coverage area and cell capacity. High data rate services will cause more interference to be generated within the cell in use than the lower data rate users. This in turn causes the cell coverage range to decrease. For example, a lower data rate user will be able to establish and maintain a connection at a greater distance from the base station than a high data rate user. In addition, when the amount of high data rate users increases within a given cell, then the capacity of that cell will decrease proportionally. Hence it is important for the planner to be aware that both high and low data rate services should be factored into the dimensional model that will be used. It should also be noted that it is necessary to map certain services to agreed data rates. To complicate the issue further, this should be based on whether the service(s) in use require a high level of QoS, such as uninterrupted service with real-time transmission capabilities (e.g. real-time video). This balance of complex requirements and task management will be analysed and performed by the software simulation tools within the actual network planning tool being used to perform this complex task. It is crucial for the planner to understand and be aware of this concept and model along with the next topic to be covered within dimensioning: the CEs.

4.1.2 Channel Element Planning

Probably the best way to think of a CE is as one connection, utilizing one spreading code, at one point in time. As the data rate of this connection will continually change, so the spreading factor will also change. This in turn will force the radio network controller (RNC) to renegotiate the spreading code for each connection involved. Therefore CEs can be considered as quantifiable, but only through the use of a planning tool, as there are many variables involved. However, ultimately the data rate is limited by the air interface, so theoretically the number of CEs required to accommodate the used bandwidth is not generally considered to be an issue, however it is useful for the planner to be aware of this. CEs need to be planned as all the associated signalling, control, and access channels relating to each specific connection must be taken into account and sufficient bandwidth reserved for them. CEs can be set sector by sector once the predicted volume of traffic has been estimated.

To perform base station CE calculations, the following criteria must be known:

a) the offered traffic in Erlangs;
b) the control channel requirements (access channels and common control channels);
c) the number of users in a soft handover scenario.

Since Erlangs have been discussed and explained in Chapter 2, suffice it to say that they are a standard measure used for quantifying telecoms traffic. The control channel requirements refer to various control channels, all of which are performing different functions with one common task: to carry the required signalling information necessary to ensure the data channels are able to function and bypass the required traffic from point A to point B. (Please note that control channels are not covered in depth in this book. For further information regarding this area please refer to *W-CDMA for UMTS* by Holma and Toskola, 2000.) Soft handovers have been reviewed in more detail in Chapters 8 and 9 as they are also taken into consideration when performing other aspects of network planning.

Calculating the number of required channels for the offered traffic can be performed by the use of Erlang tables. These tables will in addition take into account any blocking probability. Each wideband-code division multiple access (W-CDMA) carrier will require a common control channel and a pilot channel for signalling, therefore CEs will be required in, for example the access channels in the uplink and common control channels in the downlink. Pilot channels have been reviewed in Chapter 3, but can be thought of as a 'beacon'-type channel. As described earlier in this section, without any common control channels a data channel would not be able to function, hence CEs will also be required for the control channels.

The common control channels consist of the primary synchronization channel (P-SCH), the broadcast channel (BCH), and the paging channel (PCH). The number of access channels in the uplink will be mostly dependent on the anticipated traffic load within that particular cell. Network performance measurement tools can be used to determine sectors with excessive CE overheads. This will allow for corrective actions to be taken. It may be possible to reduce the common control and access channels when a user is in a coverage area comprising hierarchical cells. Hierarchical cells are covered in further detail in Section 3.8. Both soft and softer handovers will also have an effect on the number of CEs required. With soft and softer handovers the number of CEs will correspond to the number of active connections.

CE pooling or sharing between different W-CDMA carriers and also between cell sectors is always advantageous due to the trunking gain effect. Trunking gain can be thought of as the sharing of capacity. Therefore the additional number of CEs is actually less than the number of users that are currently engaged in a soft handover scenario and this is due to the Erlang-B model, as covered in Chapter 2.

4.1.2.1 Channel Element Example

CE calculations will be calculated within the planning tool being utilized, however, it is useful to understand the basics of CEs, hence a brief example is shown below. The purpose of this example is not to delve into the workings of Erlang calculations; for this reason it is assumed that the reader is knowledgeable on this subject. However, more information including Erlang calculators can be found at http://www.erlang.com/calculator/

To calculate the number of CEs for a specific base station the variables required for the calculation are shown in Table 4.1, however this will be mostly performed by the planning tool.

Table 4.1 Channel element calculation variables.

Parameters	144 kbps	8 kbps (voice)
Combined total traffic from three sectors	31.2 Erlangs	117 Erlangs
Overheads for soft handover	30%	30%
Erlangs per sector	8 Erlangs	30 Erlangs
Elements for control and pilot channels	2	2
Number of elements/sector (from Erlang table) including pooling	14	44
Total number of channel elements	16	46
Number of CEs without pooling	–	49

In reality the variety of base station configurations will vary; however, a common arrangement just as in global system for mobile communication (GSM) will be a '1 + 1 + 1' scenario, meaning one carrier per sector. Hence for a common base station configuration of three sectors this would be identified as a single cluster with a common pool of CEs. These pool or shared elements can be considered as 'floating elements' or as discussed above, common CEs which are available for sharing. With regard to CE sharing this is considered beneficial, as it will free up resources for other users.

One CE for the common control channel and the pilot channel will also be required for the downlink and likewise, an additional CE will be required for access in the uplink. In the event of a soft handover scenario occurring the number of required CEs is reduced; in certain cases, this could be less than the 30 per cent of the actual overhead traffic as pooling of the CEs will enable this reduction to be possible. Hence, pooling or sharing the resources can actually reduce the amount of CEs.

To conclude this, two-way uplink and downlink soft handover will require two CEs and hence, three-way soft handovers will require three elements.

In addition, the same CE can be used when a softer handover is occurring. It should also be noted that if macrocells are used to overlay microcells, the microcells do not necessarily require as many common control channels and access channels as the initial access traffic will be mainly handled via the macrocell. In short, if CEs can be shared then extra capacity will be available.

With regard to CE planning, detailed analysis of the interaction of coverage and Erlang capacity is a complex issue. This involves power control, soft handovers, fading, and the mobility mix of users (including the differences between forward and reverse link). The mobility mix refers to the differences in the physical movement speeds of the users that are all holding current connections, while travelling within the cell.

$$\frac{41 \text{ CEs}}{3 \text{ sectors}} = 13.7 \text{ CEs} \tag{4.1}$$

To conclude, the planner should be aware of the requirements for CE planning, even though it will be algorithmically calculated. In addition, soft connections (i.e. the users that are currently engaged in a soft handover scenario) are considered beneficial, as this can free up resources therefore resulting in a network capacity increase.

The above example, which was based on Ojanpera and Prasad's publication, '*WCDMA: Towards IP Mobility and Mobile Internet*' (2001), has illustrated how CE calculations can be performed for both the uplink and downlink, as these in turn will have an effect on capacity.

4.2 Transmission Capacity

The transmission medium that will be used within UMTS is called asynchronous transfer mode (ATM). ATM requires minimal overhead signalling and is therefore well suited for the requirements of UMTS as it is a packet-based technology. ATM can be used both in the radio access network (RAN) as well as the core network (CN) and has the ability to pass both data and voice traffic reliably, efficiently, and with the required QoS for the type of service used. ATM can be used as a transmission integration platform from 2.5G (general packet radio systems, GPRS) to 3G. This is highly beneficial as operators are able to have both flexibility and protection of investment.

Each ATM cell has two parts: a 48-byte-long payload carrying the user data and a 5-byte-long header with cell address information (see Figure 4.1). This header carries the generic flow control (GFC), the virtual path indicator (VPI), and virtual channel indicator (VCI) among others, which are all shown in the afore-mentioned diagram (Figure 4.1).

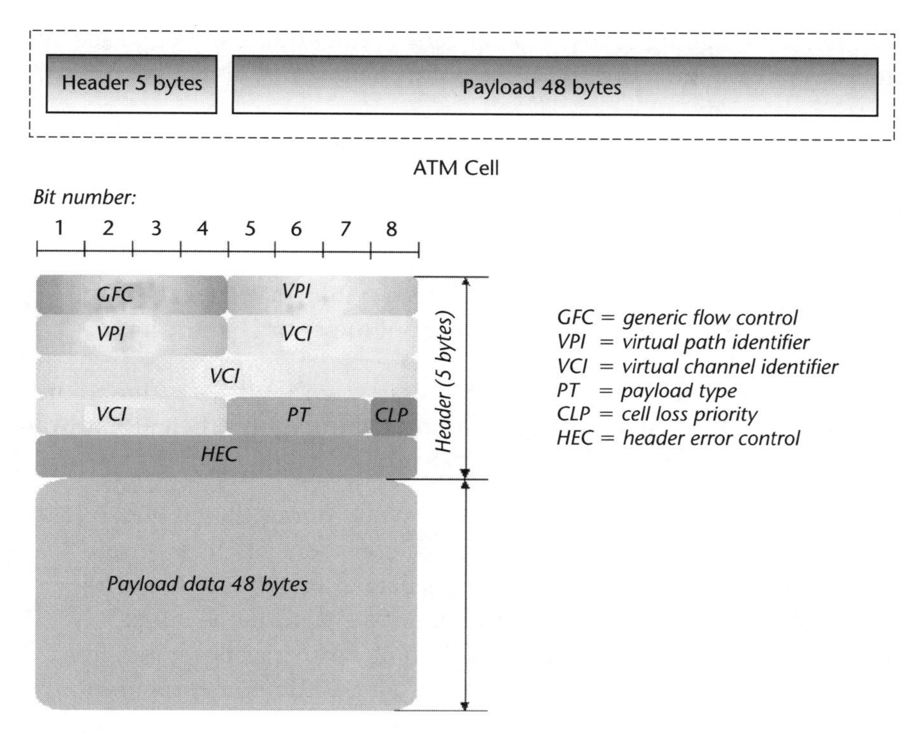

Figure 4.1 *ATM cell.*

ATM has the ability to transmit the ATM cells, and then re-assemble them at their final destination. This is achieved by dividing the upper-level data units into 53-byte cells for transmission across the physical medium. It is able to function independently, regardless of the type of transmission being generated at the upper layers, including the speed of the physical-layer medium below it.

This ensures that ATM technology is able to transport all types of transmissions (e.g. data, voice, video, etc.) in one integrated data stream across any medium. ATM uses short, fixed-length packets called cells for transport and each cell is capable of holding a maximum of 384 bits. The total amount of address information within the header consists of 40 bits. The information is then divided among these cells, transmitted, and then re-assembled at its final destination.

However, we are going to lose 9.5 per cent of total traffic in overheads, as can be seen from the following three points below:

i) A cell includes 384 information bits (payload: 48 bytes and header: 5 bytes).
ii) A cell has 40 bits of address information (contained in the header, which is 5 bytes).
iii) The required overheads (5 bytes of header information) consist of $40/(384 + 40) = 0.094 \approx 9.5$ per cent of the entire bit stream.

The advantage of ATM over other transmission mechanisms is due to the uniform size of the cell which can be routed with highly efficient hardware technology rather than the slower software.

Nonetheless, when ATM is carried over a synchronous digital hierarchy (SDH), there will be other overheads to be considered. SDH is a well-known technology used for synchronous data transmission. It can be thought of as the international equivalent of a synchronous optical transmission network. This type of technology provides faster and less expensive network interconnections than the traditional plesiochronous digital hierarchy (PDH) equipment. The term synchronous refers to the fact that all bits from one connection are carried within one transmission frame. Alternatively, plesiochronous refers to an 'almost' synchronous connection, whereby all the bits may be extracted from more than one transmission frame. The following synchronous transport modules (STM) are used in SDH and the rates are as follows:

- STM-1 (155 Mbps);
- STM-4 (622 Mbps);
- STM-16 (2.5 Gbps);
- STM-64 (10 Gbps).

The Consultative Committee for International Telegraph and Telephone (CCTTT) which is now known as the ITU-T (for Telecommunication Standardization Sector of the International Telecommunications Union), have recommendations defining a number of basic transmission rates within SDH. The first transmission rate noted consists of 155 Mbps and is referred to as an STM-1. These recommendations have also defined a multiplexing structure which allows an STM-1 signal to transport a number of lower rate signals as payload, thus allowing existing PDH signals also to be transported over a synchronous network. This is considered advantageous as it maximizes the use of the bandwidth generated by SDH.

To summarize, combining ATM and SDH and allowing ATM to emulate SDH within certain areas of the network will enable greater network efficiency to be achieved. This removes the need for SDH and therefore reduces the number of network layers, hence reducing both the costs involved and the network complexity. It is critical for the planner to be aware of the high overheads required when taking transmission lines into account for a UMTS system. There is an increasing demand for higher data transmission capacity as these high bit rates for both circuit- and packet-switched services will allow users to access mobile multimedia simultaneously in one connection. UMTS will use the same CN as both the existing GSM and GPRS systems, allowing operators to use the existing infrastructure during the roll-out phase; however, this will necessitate upgrading the transmission backbone network to allow for the higher capacities required.

4.3 Multi-user Detection

Multi-user detection (MUD) can be thought of as a method that enables the intracell interference to be reduced, thus in turn increasing the system performance and subsequently the system capacity. This is beneficial as CDMA systems are interference based, in that they operate on interference and hence, from the viewpoint of system capacity and receiver performance, they are interference limited.

Capacity increase is dependent on the radio environment, the system load, and the efficiency of the algorithm. MUD is also able to alleviate the near-far problem associated with a W-CDMA system. As shown in Figure 4.2, each user's transmitted data bits are distributed by the spreading codes and the signals transmitted over a multiple access channel. In the receiver the signal received is correlated with a reproduction of the user spreading codes. A multiplier together with an 'integrate and dump' function make up the correlation.

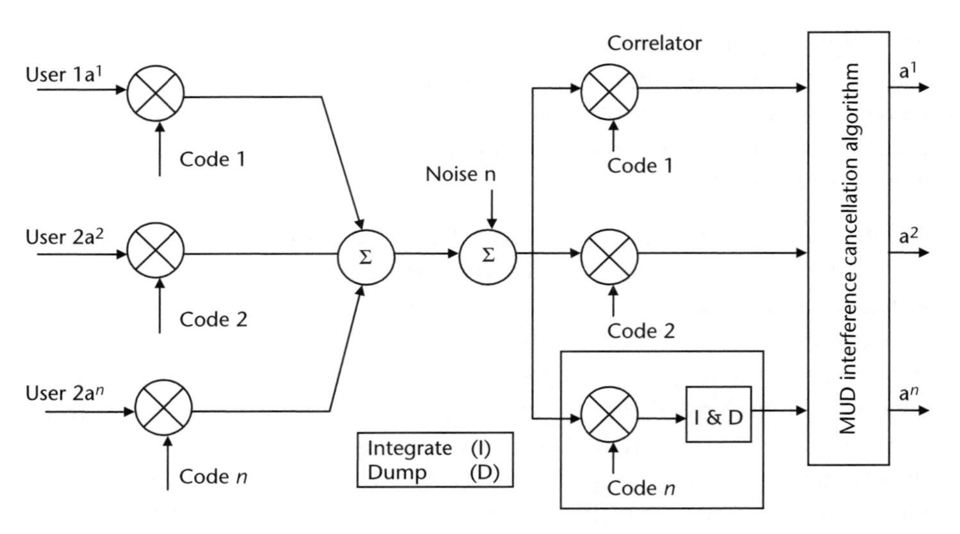

Figure 4.2 System model of MUD. Source: WCDMA: Towards IP Mobility and Mobile Internet, *Ojanpera and Prasad.*

Therefore the function of MUD is to process the signals equally from the outputs of the correlations which removes the unnecessary multiple access interference from the required signal. Hence, the actual output is an estimate of the required data bits. As there will always be multiple users, there will always be a need to have multiple MUD receivers.

However, this is considered expensive to implement and is not practical from a physical or power requirement point of view for the user equipment (UE). The efficiency of MUD will vary, dependent on different radio environments and thus introduces another variable factor to be considered when planning. This variable factor will be dependent on the channel estimate, delay estimate, detection scheme in use, and any power control errors occurring. However, it is generally considered that MUD is likely to be introduced into the base stations as an upgrade to increase performance and capacity at a later date.

MUD can be used to eliminate the interpath interference in multi-path channels even for just one user. With high data rates, interpath interference is quite substantial due to a low processing gain; also any fading occurring here can be mostly compensated by the large dynamics allowed for the uplink power control. With regard to the downlink, the UE cannot obtain enough diversity as compared to the base station.

On the positive side, the UE's transmission power can be reduced if the gain achieved from MUD in the uplink range is not used.

4.3.1 Range Extension with Multi-user Detection in Loaded Networks

The most practical method of utilizing MUD consists of implementing it within the base station, as this can be used to extend the range in a loaded network and also to increase the uplink capacity. In addition, the reduction in the range as a function of load can be, to a certain extent, advantageous. In such a case, part of the intracell interference cancellation at the base station is attributed to MUD. Another advantage of implementing MUD in the base station is to enable the coverage and range to be less sensitive to the uplink load.

MUD efficiency will however be dependent on the interference cancellation algorithm, the channel estimation algorithm, and the physical speed of the UE. The fractional loading, classed as 'F', which has to be predicted for each cell individually, will affect the performance of MUD with respect to the cell range.

A model calculation for the cell range acquired from an actual simulation can be seen in Figure 4.3 – all distances are in kilometres – and a simulated plot indicating the range as a function of efficiency of base station MUD has been illustrated in Figure 4.4(a and b). These graphs show that a base station without MUD corresponds to an efficiency of zero per cent. Depending on the efficiency of MUD, the cell range will vary considerably. As can be seen in Figure 4.4(b),

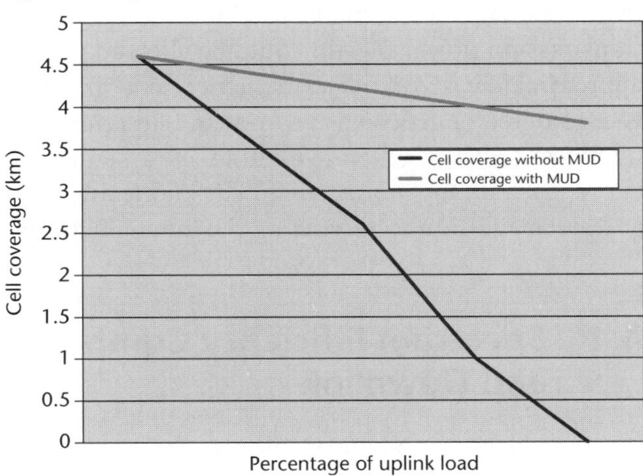

Figure 4.3 Simulated cell average with and without base station MUD.

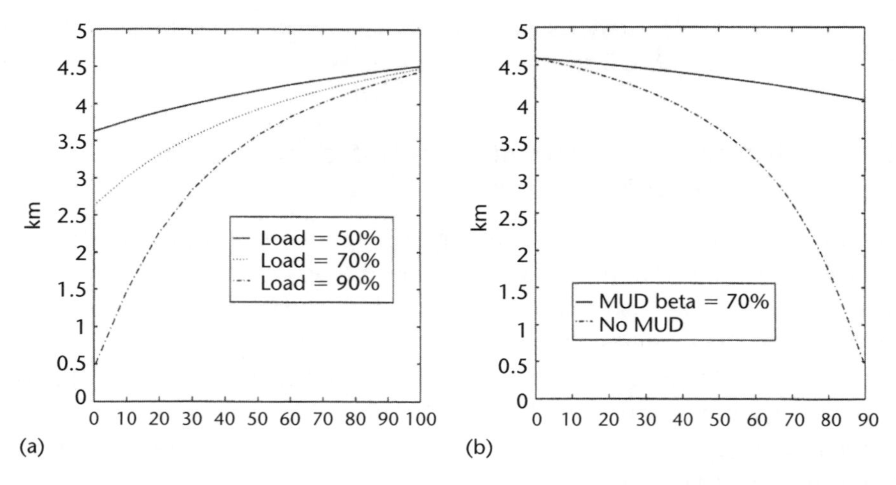

Figure 4.4 Range extension with MUD in loaded networks. (a) Efficiency of MUD, beta (%). (b) Uplink load as a percentage of maximum downlink capacity (%). Reproduced with permission from Artech House, WCDMA: Towards IP Mobility and Mobile Internet, *Ojanpera and Prasad.*

the ranges both with and without MUD are displayed. Regardless of whether the actual required coverage can be met without base station MUD, it can still be beneficial as the advanced receiver algorithms at the base station can be used to reduce the transmission power of the UEs.

An example would be to assume that MUD efficiency is 70 per cent and the fractional cell loading is also 70 per cent. A reduction in UE transmission power of 7 dB could be allowed at a 70 per cent load and approximately a 3 dB power reduction at a 50 per cent load. This reduction is considered beneficial as any power reduction will result in less noise or interference being generated within the system and consequently will allow the cell range to increase. The end result will be to allow more users access, in addition to increasing the longevity of the UE's power source.

4.4 Spectrum Efficiency Comparison – Uplink and Downlink

It is possible to use a type of software-driven simulator to predict capacity results in the network. One possibility would be to utilize a code-division test bed (CODIT: CDMA for UMTS) simulator, which, by providing

Table 4.2 Spectrum efficiency comparison.

	Uplink	Downlink
Macro 144 kbps	~2(2*x*)	*x*
Macro 12 kbps	~2*x*	*x*

Sample derived from CODIT simulator – cell capacity in kbps/cell.

Uplink: A 3–4 dB gain can be achieved with antenna diversity. The gain increases further if MUD is used in the base station.

Downlink: The UE's receiver is unable to obtain enough diversity. In soft handover the UE receives 'X' transmissions from different base stations, nor orthogonal.

simulation results for the uplink and downlink capacity, can also be used to assess spectral efficiency. An example shown in Table 4.2 indicates some sample capacity results derived from a CODIT simulator. As can be seen in Table 4.2, the uplink capacity is approximately two to four times higher than the downlink capacity if both services are taken into consideration. Taking the above into consideration if MUD, as previously discussed, is implemented it can almost double uplink capacity when compared to a standard type receiver. MUD gain can also be achieved in the downlink and is attributed to the orthogonality factor.

There is a difference between the downlink orthogonality and the MUD gain achieved in the uplink when considering soft handovers. In a soft handover state MUD gain can be achieved, as one uplink transmission is received at two base stations and hence the interference generated can be cancelled at both base stations. However, the two received signals at both base stations are not orthogonal and it also should be noted that, due to multi-path propagation especially in a macrocell, only minimal gain will be achieved.

The differences in spectrum efficiency are due to the fact that in the uplink base station antenna diversity can be used with less complexity and hence achieve a 3–4 dB gain. This antenna gain increases further if MUD is deployed in the base station receiver.

4.5 Uplink and Downlink Load Factors

4.5.1 Uplink Load Factors

Uplink load factors need to be taken into consideration as these will have an effect on capacity. Table 4.3 provides an example of the parameters used

Definitions	Suggested values/parameters
a) Eb/No required to meet predefined thermal and interference noise	– Dependent on UE speed
	– Data rate
	– RX antenna diversity
	– Multi-path fading
	– Type of service
b) Activity factor of user 'X' at physical layer	– 0.67 for speech
	– 50% speech activity and DPCCH overhead during DTX
	– 1.0 for data
c) Other cell to own cell interference ratio seen by the base station receiver	Macrocell with omni-directional antennas (55%)
d) Data rate of user 'X'	Dependent on type of service

Table 4.3 Uplink load factor calculation parameters (based on WCDMA for UMTS, *Holma and Toskola, Wiley).*

to calculate the uplink load factor. The activity factor of the user is the data rate of the user and the total received wideband power including thermal noise power in the base station. A typical speech value for the activity factor is assumed to be 0.67 (in a range of 0 to 1).

The signal energy per bit divided by the noise spectral density (Eb/No) can be obtained from link measurements and simulations, and includes the effects of soft handovers and closed-loop power control. The macro diversity combining gain relative to the single Eb/No figure can be considered as a measure of the effect of the soft handover (refer to Chapter 6, entitled Influence of Link Budgets on 3G Coverage, for further information).

The cell environment regarding the adjacent cell to own cell interference ratio is a function of cell isolation. For example, cell environments refer to the micro/macro, suburban/urban, which also include the antenna patterns such as omni-directional, and the site configurations, such as three- or six-sector configurations. The final parameter is the data rate of user and is dependent on the type of service. Taking these parameters into account it will be possible to ascertain the load factor within the uplink.

4.5.1.1 Uplink Noise Rise

Noise rise can be described as the increase in wideband interference level over the thermal noise in the base station reception. From the graph

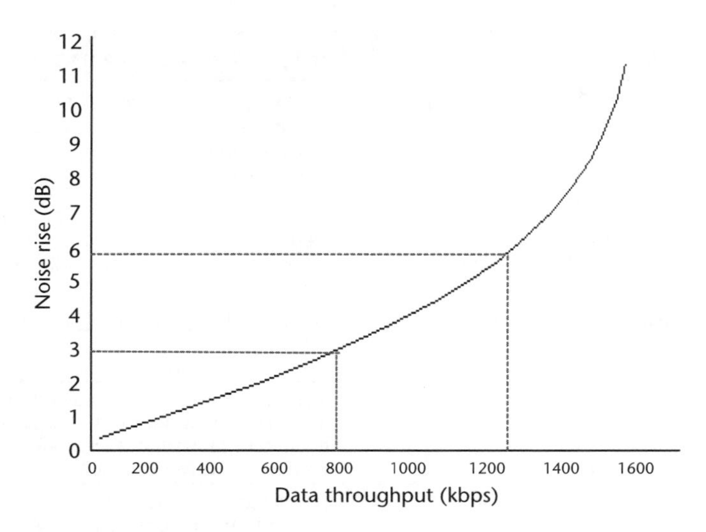

Figure 4.5 Uplink noise rise.

illustrated in Figure 4.5 it can be seen that as the noise rise increases to 3 dB, this equates to 50 per cent of the data throughput, or could be described as equivalent to a load factor of 50 per cent. However the noise rise curve is not linear, due to interference, therefore, as the noise rise reaches 6.0 dB this corresponds to a load factor of 75 per cent. The total amount of data throughput is the combined throughput of all users simultaneously. In this example, a total data throughput of 1300 kbps could be achieved with a noise rise of 6 dB. The ultimate results of this noise rise will be a reduction in cell size.

Taking the average attenuation between the base station transmitter and the UE receiver (taking into consideration the sensitivity of the UE), the minimum transmission power for each subscriber can be determined. The effect of the noise rise due to interference must be included with this minimum power level and therefore this can be stated as the required transmission power for a user situated at an 'average' location within a cell. The transmission power should be based on the average transmitted power and not the maximum transmitted power for the periphery of the cell.

4.5.2 Downlink Load Factor

As with the uplink load factors discussed in the previous section, it is these factors that are also important in the downlink. Table 4.4 lists and

Parameter classifications	Suggested dimensioning values
a) Activity factor of user 'X' at physical layer	0.67 for speech, assumed 50% voice activity and DPCCH overhead during discontinuous TX 1.0 for data
b) Data rate of user 'X'	Dependent on type of service
c) Eb/No required to meet predefined QOS (e.g. BER). Noise includes both thermal noise and interference	Dependent on UE speed, multi-path fading, type of service, TX antenna diversity
d) Orthogonality of channel user	Dependent on multi-path propagation 1 = fully orthogonal 0 = no orthogonality
e) Ratio of other cell to own base station power, received by user 'X'	Dependent on the UE's location in the cell and log-normal shadowing, each user sees a different ratio of adjacent cell to own cell BS power received by the user
f) Average orthogonality factor in cell	ITU Vehicular A channel ~60% ITU Pedestrian A channel ~90%
g) Average ratio of adjacent cell to own cell BS power received by user. Own cell interference is wideband	Macrocell with omni-directional antennas

Table 4.4 Downlink load factors (based on WCDMA for UMTS, *Holma and Toskola, Wiley).*

describes the various parameters and their suggested dimensioning values which are discussed in detail below.

The activity factor can generally be thought of as the amount of voice traffic compared to data traffic. The ratio between voice and data will become more accurate to predict following the ongoing collection of data provided from the functioning UMTS service. One important new parameter to be taken into consideration when attempting to make comparisons between the downlink and uplink load factors, is the orthogonality factor of the user.

Orthogonality can be best defined as the separational properties existing between two users. If both the users possess good separational properties they will not interfere with each other.

If the downlink orthogonal codes are used to ensure good user separation, in the case where no multi-path propagation exists, the orthogonality remains intact when the UE receives the signal in the downlink. However, if enough delay spread occurs in the radio channel, this will be interpreted by the UE as multiple access interference.

An orthogonality factor of '1' equates to perfectly orthogonal users. Within multi-path channels typical orthogonality factor values are between 0.4 and 0.9 (in a range of 0–1).

4.5.3 Soft Handover

A soft handover overhead can be defined as the total number of active connections, divided by the total number of users minus one.

The gain achieved by soft handovers relative to the single link Eb/No must be taken into consideration and is known as the macro diversity combining gain. This is measured as the reduction in the required Eb/No for each user.

This gain can be calculated by performing link-level simulation analysis.

If a user is currently engaged in a soft handover scenario, then all the other base stations in the active set are included or counted as 'part of the other cell'. The maximum power that the base station amplifier can provide is the limiting factor in the downlink. When the network approaches saturation, a similarity exists between both the uplink and downlink load factors, since, upon reaching pole capacity, the noise rise over the thermal noise increases to infinity. In this case an estimate of the total base station transmission power is required to enable the downlink dimensioning to be performed.

4.6 Uplink and Downlink Capacity

4.6.1 Uplink Capacity

Following the launch of a network in rural areas, capacity is of secondary importance compared to the coverage required. However, in an urban environment, where a high data rate service is being provided by the standard outdoor base stations then coverage can be reduced due to the localized capacity requirements. In contrast when capacity is not in demand the cell size and therefore coverage increases.

As the power output of the UE is always lower than the output of a base station, the macrocell coverage can be calculated by the uplink range. Hence, the output power of a macrocell base station is around

40–46 dBm (10–40 W) per sector and the output power of the UE is typically 21 dBm (125 mW).

Using a standard propagation model, such as Okamura–Hata's (which is one of the models used in the planning tool), the improvements made in the link budget on the corresponding cell radius can be calculated. The results can be seen in Table 4.5, where there is a distinct relationship between the cell coverage area and the base station density. For example, with a link performance improvement of 2 dB from 4 to 6 dB there is an improvement in the required cell density from 59 to 46 per cent.

Base station density – relative number of sites (%)	Link budget improvement (dB)
27	10
46	6
52	5
59	4
68	3
77	2
88	1
100	0 – Reference case

Table 4.5 Uplink coverage.

4.6.2 Downlink Capacity

In the downlink the air interface is limited by interference and this should be taken into consideration, therefore the amount of interference and the cell capacity will need to be estimated. In addition the evaluation of the effects of the orthogonal codes on the downlink capacity in both the micro- and macrocell environments should also be taken into account. The effect of intercell interference, produced by adjacent base stations, will have a marked effect on the capacity within the downlink. However, the use of orthogonal codes will ensure that the effect of the intercell interference from adjacent base stations is reduced in the downlink as compared to the uplink. In addition, the network planning and the propagation environment will have an effect on the amount of interference from the adjacent cells if the network plan has not been performed correctly.

In dense urban environments, the street corners tend to isolate cells more than the macrocells, so using microcells will ensure that in such an environment the intercell interference will be lower than in a macrocell environment. This cell isolation is represented by the 'adjacent to own cell interference ratio' parameter. A greater orthogonality for the downlink codes is likely to exist in cell environments, as generally there is less multi-path propagation present in microcells. Additionally, if less multi-path propagation exists, this will result in less multi-path diversity, therefore a higher Eb/No requirement in the downlink can be assumed in microcells as opposed to macrocells.

As illustrated in Table 4.6, the assumed loading is 80 per cent in the downlink and 60 per cent in the uplink. As coverage is generally more difficult to obtain in respect to an uplink, a lower loading in the uplink is assumed.

	Microcell	Macrocell
Adjacent to own cell interference ratio	0.20	0.65
Uplink Eb/No (dB)	1.50	1.50
Uplink loading (%)	60.00	60.00
Downlink Eb/No (dB)	8.00	5.50
Downlink loading (%)	80.00	80.00
Downlink orthogonality	0.95	0.60

Table 4.6 Assumed throughput calculations.

As previously discussed in this chapter, a higher loading will cause a reduction in coverage. Receive antenna diversity is assumed for the uplink, however, the effect of transmit antenna diversity is not included for the assumed downlink performance. The example data throughput results shown in Table 4.7 have been calculated for both the uplink and the downlink and are taken from the assumptions of the derived calculations previously shown in Table 4.6. If a frame error rate (FER) of 10 per cent is assumed then the data throughput will equate to 90 per cent of the calculated values. The FER can be thought of as the performance measurement rate of a network that is transmitting data in packets, as the received packets will arrive in frames and must be decoded and reconstructed to acquire the wanted signal. These frames of data often become corrupted due to the fact that the air interface is the weakest link and is the least robust method of transmission along the communication path. Hence, as the FER increases, the network performance decreases. Therefore this figure of 90 per cent is derived on the assumption that an FER of 10 per cent will occur. With microcells there is a fairly reasonable balance

	Microcell	Macrocell
Uplink (Kbps)	1430	1040
Downlink (Kbps)	1440	660

Table 4.7 Data throughput/sector/carrier in macro- and microcell environment.

between the uplink and downlink capacities. However, in macrocells the downlink throughput is lower than the uplink throughput. Due to the use of orthogonal codes, the downlink capacity is more dependent on both the multi-path and propagation environments than the uplink. Cell capacities on average will be greater if the UEs are located closer to the base stations. However, for this example, in Table 4.6 it is assumed that all users are distributed uniformly over the cell area. It will not be possible to support 2 Mbps per user in every cell if the users requiring 2 Mbps are scattered about independently within the cell area, including the cell periphery.

As we have discovered above, some of the main issues affecting the downlink capacity are orthogonality of the users and cell loading, and these will vary depending on the type of cell in use. Some assumed examples have been given and hence the planner should be aware of all the issues with regard to downlink capacity. The following section deals with how to take full advantage of capacity with a series of potential improvements.

4.7 Capacity Improvements

There are a number of methods available to improve capacity such as adding additional frequencies, sectorizing, utilizing transmit diversity, and using lower bit rate codes. These subjects are covered in this section, starting with additional frequencies.

4.7.1 Additional Frequencies

Additional frequencies are useful as they will enable greater capacity within a cell. For example, if an operator has purchased more than one block of the spectrum, a number of carriers can be used to balance out the traffic loading and also boost the capacity per cell. One method to alleviate any further investment in expensive linear power amplifiers is to share one power amplifier between a number of carriers. Furthermore the most efficient method of power amplification is achieved by sharing one power amplifier between two carriers, as the loading curve can then be divided between the two carriers. In addition, when the load curve starts to diminish, the required transmission power for each user is reduced. However, increasing the downlink transmission power is likely to achieve only a minimal capacity gain from the load curve. Hence, this is not considered to be an efficient method for increasing the downlink capacity.

4.7.2 Sectorization

A standard method of increasing the capacity of a site can be performed by sectorization. Ideally 'y' sectors should give 'y' times higher capacity, however, in reality efficiency is usually around 90 per cent. The downside of sectorization as far as upgrading capacity is concerned, is that, by increasing the number of sectors, new antennas will have to be installed and the radio plan must be reviewed then optimized. It is inevitable this will occur as traffic demand increases, so upgrading from an omni-site to a three-sector site will give an increase in capacity of around 2.7, and hence for a six-sector site a capacity increase of around 5.5 can be expected. Increasing the number of sectors will also increase the antenna gain, thus improving the coverage, but this approach is limited by potentially escalating costs related to the increase in sectors and will cause further planning issues to be resolved.

4.7.3 Transmit Diversity

To improve performance with downlink transmit diversity (refer to Chapter 6, for more information on transmit diversity) the data can be split into two separate data streams and then spread using orthogonal sequences. The multi-path diversity present within the specific radio environment will have an effect on the achieved gains. For example, if less multi-path diversity is present, then the capacity gains in the downlink will be higher when transmit diversity is employed. With this in mind, the highest capacity gains are likely to occur in micro and picocells, where multi-path diversity will be limited.

4.7.4 Lower Bit Rate Codes

Finally, it is possible to increase voice capacity with adaptive-multi-rate (AMR) speech codes. AMR speech codes have eight source rates available and extra capacity can be achieved by using a lower source rate. The AMR codes will allow a trade-off between speech capacity and quality as required. With AMR the number of connections can be increased, while simultaneously reducing the data rate per each user.

This concludes the topic on *network dimensioning*. The issues covered here should enable the planner to be aware of the requirements necessary to ensure the UMTS network is correctly dimensioned. Issues such as traffic modelling, transmission capacity, spectrum efficiency, and both uplink and downlink load factors have been covered and will all in turn have an effect on the capacity of the network.

Summary for Part II

This concludes Part II which focused on the capacity parameter of the CCQ model. We have dealt with such issues as detailed network planning, capacity and soft handover planning, capacity enhancements, and, to complete the capacity issues, network dimensioning. We have also discussed traffic modelling, CEs, MUD, uplink and downlink capacity, soft handovers, and finally some options for improving capacity.

This leads into the next major element of the CCQ model, that of *coverage and network planning*.

PART THREE

Coverage and Network Planning

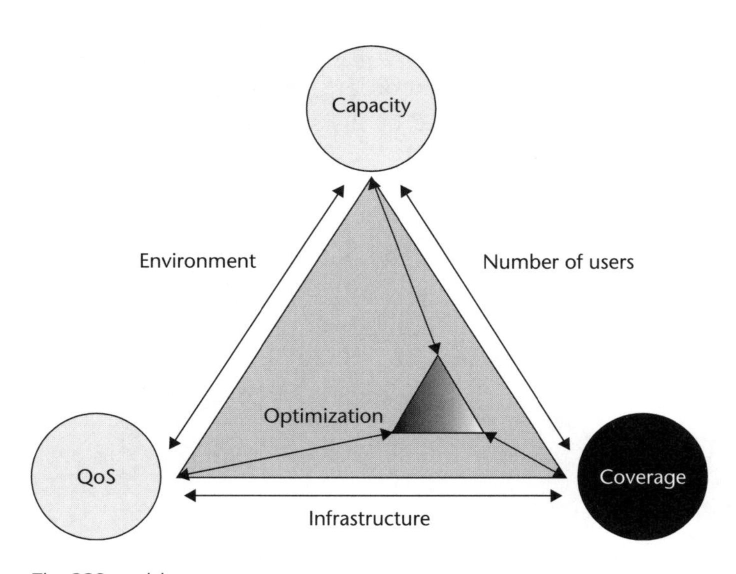

The CCQ model – coverage.

Introduction

In this section both second-generation (2G) and third-generation (3G) co-existence is discussed, along with the associated issues and parameters that need to be considered to ensure the optimal coverage, capacity, and quality of service (CCQ) can be achieved with respect to both 2G and 3G. From a coverage perspective, it is inevitable that a 3G system will not be able to offer the same widespread coverage as the existing global system for mobile communication (GSM) (2G) networks and the reasons for this will become clear in this section.

Also discussed are both the uplink and downlink budgets, including various working examples. It is these budgets that will have a marked effect on the network's coverage and so all attempts to maximize these must be achieved in order to attain the best possible coverage.

It will also be necessary to perform optimization for both voice and packet services and in addition be able to support higher rate circuit-switched services.

To achieve a smooth, backwards compatible evolution from an existing 2G system, eventually dual-mode terminals will be available and these will be introduced into the network in order to allow gradual build-up of high data rate services in areas covered by 2G services.

CHAPTER FIVE

3G Co-planning and Co-existence

This section refers to the coverage aspect of the coverage, capacity, and quality-of-service (QoS), abbreviated as CCQ model and is the third significant factor, leading to optimization in finalizing the required coverage (refer to figure in Part III). Therefore, this chapter starts by covering the crucial co-location issues with regard to both third-generation (3G) planning and network roll-out, as ultimately each operator will be forced to re-use as many of their existing second-generation (2G) sites as possible. Initially only areas where high data traffic usage is predicted will be covered. In addition, there will be a need for multi-mode terminals as they will be essential in offering both 2G and 3G services simultaneously. The use of existing sites and the sharing of the same infrastructure, coupled with a reduced implementation roll-out time frame, will result in high cost savings. However, there are numerous problems associated with co-existence, therefore the topics covered in this chapter provide the reader with the knowledge regarding the major issues and pitfalls to be taken into consideration when co-planning for a universal telephony mobile system (UMTS) network. These topics include intermodulation, isolation, and adjacent channel interference (ACI). In addition, uplink calculations and uplink simulations are explained along with UMTS carriers, frequency sharing, and guard bands.

5.1 Co-existing with 2G

One of the most important considerations when rolling out a 3G network is how it will co-exist with an existing global system for mobile communications (GSM) system. For established operators the most viable strategy is to initially plan for local coverage of 3G, thus using the existing 2G system to provide stable, circuit-switched, low data rate coverage. One method to alleviate any possible congestion in the existing legacy network would be to make use of the greater spectrum efficiency available within wideband-code division multiple access (W-CDMA). This is somewhat complex to perform, and requires the initial introduction of dual-mode terminals with wideband capabilities into the network. The old carriers will then need to be cleared to ensure that any loss of service is avoided, so that the new carriers can be utilized simultaneously. Multi-mode terminals will play a large role in offering data services of the future, and once data usage becomes more commonplace, then it is likely that single mode data terminals will evolve. To ensure implementation and co-existence with different air interfaces, the following seven items, which are the most critical factors when combining a 3G network with an existing 2G network, must be carefully considered and are covered in further detail throughout this book:

i) Co-planning (deployment scenarios) (see Chapter 5 and Table 5.1).
ii) Intermodulation (see Section 5.3 and Figure 5.1).
iii) Isolation (see Section 5.4 and Table 5.2).
iv) External operator interference (see Section 5.5 and Figure 5.2).
v) ACI (see Section 5.5 and Figure 5.2).
vi) Inter-system handovers (slotted mode) (see Section 5.8 and Figure 5.10).
vii) Guard zones and bands (see Section 5.7.2 and Figure 5.9).

The final aspect of 2G co-existence concerns GSM equipment manufactured before 1999. Filtering will be necessary to ensure compatibility can be maintained and this will also ensure that interference levels can be minimized to and from the older types of GSM equipment. This is explained in more detail in Section 5.4 in this chapter.

At the time of publication, a proportion of the existing European operators have already embarked upon and are continuing their 3G base station roll-out, whereby a high percentage of their existing 2G sites are being utilized.

To summarize, as 2G will continue to generate good revenue streams, it can be coupled with the ability to provide 3G users with that ever-important 'network connection', whereby once the user who possesses a multi-mode terminal travels out of a 3G coverage area, the terminal will automatically search, and 'log-on' to the existing 2G network. This is the current strategy likely to be employed by most existing operators. With regard to capacity, careful planning must be implemented, in addition to planning the transmission backhaul infrastructure.

5.2 3G Co-planning with 2G Networks

Site sharing will be more prevalent within UMTS (3G), due to the increasing problems associated with site acquisition. One such problem arises from public sensitivity to the yet unproven dangers of radio frequencies (RFs) and hence, procuring suitable sites will be particularly challenging. In addition, within dense urban areas a high volume of sites will be required, thus every possibility to site share with the existing 2G network must be explored.

Table 5.1 shows the assumptions indicating the pathloss results as a comparison between 3G and 2G coverage (please also refer to the additional notes for this table at the end of this section). The user equipment (UE) transmission power is assumed to be 21 dB (0.125 W) for 3G and the fast-fading margin in this example includes any gains achieved by soft handovers. The fast-fading margin for W-CDMA includes the macro diversity gain against fast fading. It can be seen here that the pathloss is identical in this sample for both the 2G speech service and the 144 kbps 3G data service. It should be noted that W-CDMA sensitivity assumes a 4.0 dB base station noise figure and an energy-per-bit/noise-per-spectral density (Eb/No) of 5.0 dB for 12.2 kbps speech, 1.5 dB for 144 kbps, and 1.0 dB for 384 kbps data. A fair average for GSM sensitivity, based on manufacturers' specifications, is around −110 dB with receive antenna diversity. As indicated in the example shown in Table 5.1, a specified coverage area being used for a GSM network would be the same coverage service area for a 144 kbps data service. Taking into account the old 2G GSM 900 sites, if co-location is being considered for UMTS coverage requiring data rates up to 144 kbps, then a 10 dB coverage improvement will be required to compensate for the higher pathloss. The reason for this can be seen when comparing the parameters for GSM 900 and 3G 144 kbps (as shown in Table 5.1). An interference margin of 1.0 dB is reserved for

				Actual	
Assumptions	**3G-384 kbps**	**3G-144 kbps**	**3G-speech**	**2G-1800 speech**	**2G-900 speech**
RX sensitivity	−113 dBm	−117 dBm	−124 dB	−110 dBm	−110 dBm
Interference margin	2 dB	2 dB	2 dB	0 dB	1 dB
UE TX power	21 dBm	21 dBm	21 dBm	30 dBm	33 dBm
Fast-fading margin	2 dB	2 dB	2 dB	2 dB	2 dB
Body loss	–	–	3 dB	3 dB	3 dB
Base station antenna gain	18 dBi	18 dBi	18 dBi	18 dBi	16 dBi
UE antenna gain	2 dBi	2 dBi	0 dBi	0 dBi	0 dBi
Relative gain from lower frequency compared to 3G frequency	–	–	–	1 dB	11 dB
Maximum pathloss	150 dB	154 dB	156 dB	154 dB	164 dB

Table 5.1 2G co-planning (reproduced by permission of Wiley & Sons: WCDMA for UMTS, *Holma and Toskola).*

GSM 900 because the small amount of spectrum in 900 MHz does not allow large re-use factors. Finally, this sample comparison makes the assumption that these older networks are planned in a coverage limited manner, due to the average cell radii being larger when using the 900 MHz band. Coverage limited means that the cell size could be larger from a capacity point of view. Nevertheless, GSM 900 cell sizes are smaller in dense urban areas to ensure the required capacity is available.

With regard to the antennas in this sample table, the antenna gain assumes three-sector configurations in both GSM and W-CDMA, and a 2.0 dBi antenna gain is assumed for the data terminal (dBi refers to a theoretical isotropic antenna gain).

Finally, the attenuation in GSM 900 MHz is assumed to be 110.0 dB lower than in the UMTS band and in the GSM band 1.0 dB lower than in the UMTS band.

To conclude, the pathlosses shown in Table 5.1 should provide the reader with a typical example of what to expect when performing co-planning scenarios. The subject of pathloss itself has been covered in more detail in Chapter 9.

5.2.1 Co-planning Parameters

The following notes (taken from Holma, H. and Toskala, A. (2000). *WCDMA for UMTS Radio Access* reproduced by permission of Wiley publications) relate to the parameters in Table 5.1:

- Sensitivity (W-CDMA) assumes a 4.0 dB base station noise figure and a Eb/No of 5.0 dB for 12.2 kbps speech, 1.5 dB for 144 kbps and 1.0 dB for 384 kbps data. GSM sensitivity is assumed to be −110 dB with receive antenna diversity.
- The interference margin corresponds to 37 per cent loading of pole capacity. An interference margin of 1.0 dB is reserved for GSM 900 as the small amount of spectrum in 900 MHz does not allow large re-use factors.
- The fast-fading margin for W-CDMA includes the macro diversity gain against fast fading.
- The antenna gain assumes three-sector configuration in both GSM and W-CDMA.
- Body loss accounts for the loss when the terminal is close to the users head.
- A 2.0 dBi antenna gain is assumed for the data only terminal.
- The attenuation in 900 MHz is assumed to be 11.0 dB lower than in the UMTS band, and in the GSM 1800 band, 1.0 dB lower than in the UMTS band.

5.3 Intermodulation

Intermodulation can be described as the frequencies of signals that are generated as a result of the combination of harmonic generation and mixing action of multiple input signals by non-linear impedances in a system or device. Intermodulation in a W-CDMA system can occur if there are two or more different frequencies present at the input of any amplifier, which will cause unwanted intermodulation products to be generated.

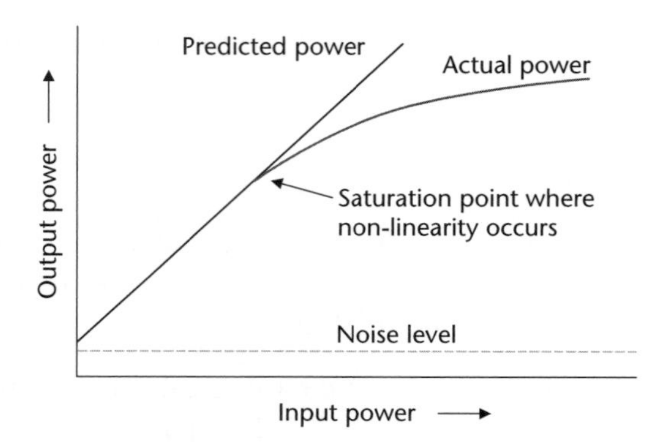

Figure 5.1 Intermodulation.

This can be better explained by referring to Figure 5.1. As can be seen at the saturation point, this is when the signal becomes distorted and hence, becomes non-linear. In addition, the relation to the power of the original signal can be seen, and it is at this point that unwanted harmonics and intermodulation products are created.

5.3.1 Intermodulation Products

Intermodulation can be categorized into handset or UE generated intermodulation and transmit intermodulation.

Transmit intermodulation occurs when multiple carriers use the same amplifier, thus creating intermodulation products that may spill over into the wideband spectrum causing unwanted interference, which usually occurs in the base station transmitter. Intermodulation generated by a UE will be dependent on how close the UE is to an interfering base station.

There are two intermodulation products that need to be considered, known as the second order and third order intermodulation products. The majority of quality receivers have front-end filters which will substantially reduce the second order intermodulation product. This means that the third order intermodulation problems are more prevalent. These third order intermodulation products are caused by non-linear active stages such as the low-noise-amplifier (LNA) and intermediate frequency (IF) amplifier in the receiver front-end.

This could be resolved by bypassing the LNA, which would cause a reduction in any unwanted intermodulation signals at the receiver's front-end. Problems associated with this solution include the potential for dropped calls, because the required signal would already be too weak.

A situation could occur where the third order intermodulation product could cause a dropped call. The power of the third order intermodulation products increases phenomenally as the RF gain increases, consequently the unwanted third order products quickly 'swamp' the desired signal, causing the call to drop.

To further understand intermodulation, it is best thought of as the case where a single frequency f1 is inputted through a component of an amplifier such that the output from this component is not a linear function of its input. This causes the common phenomenon of harmonics to be generated, for example f2, f3, f5, etc., as no RF component is perfect. Therefore, if two separate frequencies are present in a non-linear device, both sum and difference frequencies are produced as well as their harmonics. This is the equivalent to the result of the multiplication process of these two frequencies and is known as intermodulation products. Hence, products of f1 + f2 and f2 − f1 will be generated. As there are harmonics generated for both of these products, and these will be the sum and the difference between both the harmonics and the frequencies themselves; these are known as the intermodulation products. Intermodulation products can be considered as infinite, and exist at a low level, therefore they do represent a problem. When considering defining the orders of intermodulation products, we have to count frequencies. If we are dealing with two frequencies, then to define the order, the harmonic multiplying constants of both the input frequencies are added. Hence, f1 + f2 would produce the second order product, 2f1 − f2 would produce the third order product, and likewise 3f1 − f2 and so on. In summary, intermodulation and its products are factors that must be taken into consideration when defining the performance of both a receiver and transmitter.

5.3.2 Intermodulation and 2G/3G Site Share

When site sharing 3G sites with the existing 2G sites, assuming that the same frequency band can be used as that of the already deployed network, the possibility exists to map both the new and existing cells. Two techniques are possible here, 'one to one' mapping, and 'one to *n*' mapping. Generally, 'one to one' mapping is considered preferable, due to the fact that intermodulation problems can be minimized. This is because with 'one to one' mapping, generally the transmission powers radiating from both systems will be virtually similar, thus reducing the possibility for intermodulation to occur. In addition, with 'one to one' mapping both the actual cell planning and implementation will be simplified, as the cell size and coverage area will be virtually identical for both systems. 3G coverage will expand as the capacity requirement increases, so it is reasonable to assume that eventually 3G cell sizes will be the same as those cells in existing systems within certain areas. With 'one to *n*' mapping, the idea is that one 3G cell will offer a greater coverage area than a corresponding cell within the existing 2G system. This will ensure that there will be a significant difference between both systems' transmission powers, therefore intermodulation problems are far more likely to become prevalent. In the early stages it is expected that the traffic load within 3G will be rather low, thus a coverage limited deployment strategy would be beneficial. Careful planning must be performed when using 'one to *n*' mapping, because when the UE is located at the cell periphery and has reached its sensitivity limit, the higher intermodulation products may cause a dropped call to occur. The specification requirements are shown in Table 5.2 for both blocking isolation and intermodulation products, which also includes taking into account any spurious emissions that may be present. Spurious emissions can be defined as signals being transmitted on a frequency other than that for which the transmitter was designed. If adhered to, the requirements listed in Table 5.2 will significantly reduce any intermodulation and spurious noise problems between the two systems.

Specification requirements	UMTS TX to UMTS RX (dB)	UMTS TX to GSM 1800 RX (dB)	UMTS TX to GSM 900 RX (dB)	GSM 900/ GSM 1800 to UMTS RX
Intermodulation products and spurious emissions	39	35	35	90 (40 dB for BTS release 99 or later)
Blocking isolation	63	49	40	58

Table 5.2 Typical isolation requirements.

In summary, intermodulation should always be taken into consideration, especially when network planning with respect to co-locating 3G with 2G. This leads into another related issue, that of isolation.

5.4 Isolation Requirements

The isolation requirements between UMTS and GSM systems can be derived from the UMTS and GSM specifications given by the European Telecommunications Systems Institute (ETSI) and the 3G partnership project (3GPP). The specifications have been agreed by both bodies to ensure compatibility and inter-operability. Each vendor should be able to provide information which can be used to improve the isolation requirements and therefore the defining specifications. In many cases equipment manufacturers will ensure their equipment performance will exceed the isolation specification requirements.

These isolation requirements will affect the choice of antenna configuration and the filtering for both receivers and transmitters in the GSM and UMTS base stations. To define isolation, it is the actual attenuation from the output port of the base station transmitter to the input port of the base station receiver. Table 5.2 indicates the isolation requirements between 2G and 3G. It can be seen in Table 5.2 that the highest blocking isolation requirement exists between the UMTS base station transmitter and its receiver. It should also be noted that pre-1999 GSM equipment may cause or create unwanted interference with or by the 3G base stations, so in this case filtering will be a requirement.

Filtering will assist in providing better isolation with varying degrees of effectiveness between the different equipment manufacturers.

To summarize, good isolation will assist in reducing interference levels, which in turn will ensure the required coverage can be achieved.

5.5 External Operator Interference

External operator interference refers to interference generated by other 2G and 3G network providers. This is an issue that will need to be carefully considered by all operators who plan to use the spectrum allocated for UMTS to ensure the required coverage can be achieved and maintained. This can also be classed as ACI and is an inherent problem within wideband systems such as UMTS, in which large guard bands are not possible. Refer to Section 5.7.2 in this chapter for more information relating to guard bands. Due to a relatively large system bandwidth, valuable spectrum is wasted with large guard bands. These guard bands are necessary to ensure adequate isolation between adjacent frequencies, in particular other W-CDMA carriers. One method to reduce the ACI would be to use a tight spectrum mask on the transmitter and increase the receiver selectivity. However, these are quite challenging requirements for a 3G base station, particularly for a micro- or picocell base station, due to the lack of frequency spacing available between carriers. The ratio of the transmission power to the power measured after a receiver filter in the adjacent channel(s) is known as the adjacent channel interference power ratio (ACIR). Inadequate receiver filtering coupled with poor transmitter filtering creates ACI. The UE's performance degrades both the downlink and uplink performance of the adjacent channel. Power amplifier design is severely constrained by both power and size requirements in the UE which reduce the linearity of the amplifier and hence cause adjacent channel leakage (ACL) power. This is the main cause of ACI in the uplink.

Selectivity can be considered as one of the important parameters of any receiver, as this determines whether the receiver is able to acquire the required signal among the many other unwanted signals. The ACI is caused by the UE's inadequate receiver selectivity. Filtering is used to achieve good selectivity and enables the receiver to select and remove these unwanted signals. Although filtering will remove some of the unwanted interference, an amount of attenuation will be required to ensure that all ACI components are unable to cause problems. Table 5.3 shows the attenuation levels that will be required to ensure ACI can be avoided and to achieve satisfactory adjacent channel performance.

Attenuation required level (dB)	Frequency separation (MHz)
33	Adjacent carrier, 5
43	Second adjacent carrier, 10

Table 5.3 Attenuation levels. Source: WCDMA for UMTS, *Holma and Toskola, Wiley.*

5.5.1 Example of External Operator Interference

The following example illustrates the difficulties of external operator interference with regard to the coverage aspects of the network plan. This rather problematic interference scenario which could easily affect the network performance has been illustrated in Figure 5.2. Here there are

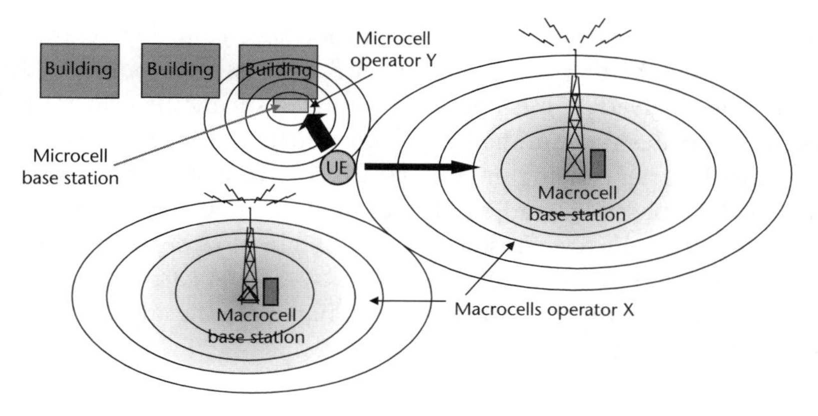

Figure 5.2 External operator interference.

large macrocells belonging to operator X and a small microcell belonging to operator Y. A UE assigned to operator X is located next to a microcell belonging to operator Y. The UE is currently maintaining a connection to a macrocell belonging to operator X. For this example both operators X and Y are operating on adjacent frequencies, so consequently part of the UE's transmissions are spilling over onto the adjacent carrier, causing interference in the microcell belonging to operator Y. The UE of operator X will experience ACI in the downlink from the microcell of operator Y. Interference will be received from the adjacent carrier due to the poor selectivity of the UE. This could result in the UE generating interference to affect all the uplink connections of the entire cell. A solution to this type of problem would be the availability of more spectrum, however this is a restricted resource. Another possibility would be to perform an inter-frequency (hard) handover when ACI is experienced in the downlink from another base station. This requires jumping to another frequency and is explained in further detail in Section 5.8 (slotted mode). To conclude, the planner should be aware of external operator interference while keeping all ACI issues in mind, in order to maintain the required coverage.

5.5.2 Adjacent Channel Leakage Ratio

As previously explained, interference between different UMTS carriers will need to be taken into consideration, therefore the selecting of each required W-CDMA carrier within the available spectrum. Hence, the ultimate aim will be to 'space' each carrier as accurately as possible, thus ensuring the received ACL is negligible, while ensuring the transmitted carrier only creates the minimal ACL with respect to the other carriers. The nominal carrier spacing in W-CDMA is 5 MHz, and with a 200 kHz raster it will be possible to adjust the centre frequency up or down in steps of 200 kHz (the term 'raster' can be thought of as describing an incremental step in frequency). Moving the W-CDMA carrier in steps of 200 kHz away from the preceding carrier will reduce the adjacent channel leakage ratio (ACLR).

The amount of transmitted power from the first carrier that 'leaks' into the neighbouring carrier is defined as the ACLR and can be best understood by the diagram in Figure 5.3. The remainder of the channel contains the required 'usable' signal. This is again illustrated in Figure 5.3 for both ACLR '*a*' and ACLR '*b*' with respect to both the first and second neighbouring carriers. However, as shown in the example in Figure 5.3, the ACLR values for the UE power classes are assumed to be 33 dB for ACLR '*a*' and 43 dB for ACLR '*b*'. Less leakage and improved linearity in the carrier improves the usable signal and can be achieved by improving the efficiency of the power amplifier. The disadvantage is that the increased linearity tends to reduce the amplifier's efficiency.

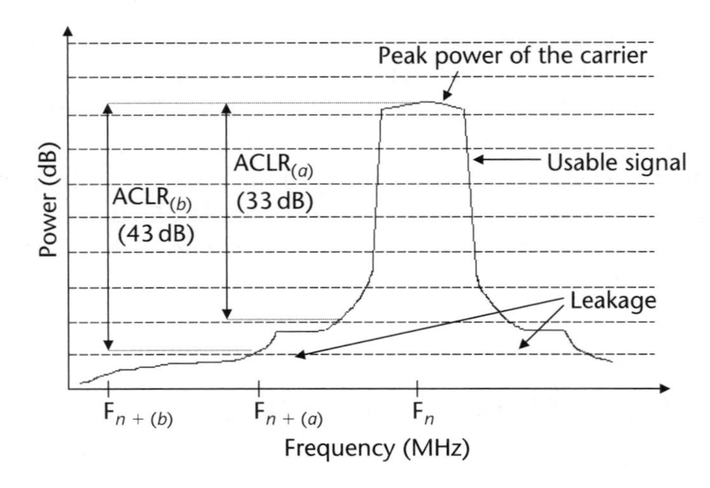

Figure 5.3 Adjacent channel leakage ratio (ACLR).

In addition, the planner should be aware that frequency accuracy requirements should be considered and must be tighter for the base stations. These should be $\mp$ 0.05 ppm, due to the fact that the baseband timing must correlate with the same timing reference for the base station carrier. Accuracy here is paramount, as this will be used by the UE as a frequency reference. The accuracy is not as critical for the UE and will be $\mp$ 0.1 ppm when compared to the received carrier frequency. As the UE moves, a frequency tolerance error should also be considered. In addition, base station frequency tolerance errors will also exist. However, it should be noted that the UE must be able to 'view' these frequency ranges that fall out of the agreed tolerance parameters. In summary, with minimal ACLR, the required coverage should be able to be adequately maintained.

5.5.3 Adjacent Channel Attenuation

Adjacent channel attenuation should also be taken into account, with regard to channel interference, as this involves reducing the power levels of the adjacent channel, thus in turn reducing the ACI. As can be seen in Figure 5.4 the more adjacent channel attenuation that is present, then, due to the non-linearity of the power amplifier, the greater the reduction in spectrum efficiency. Hence, an interfering user's output power can be

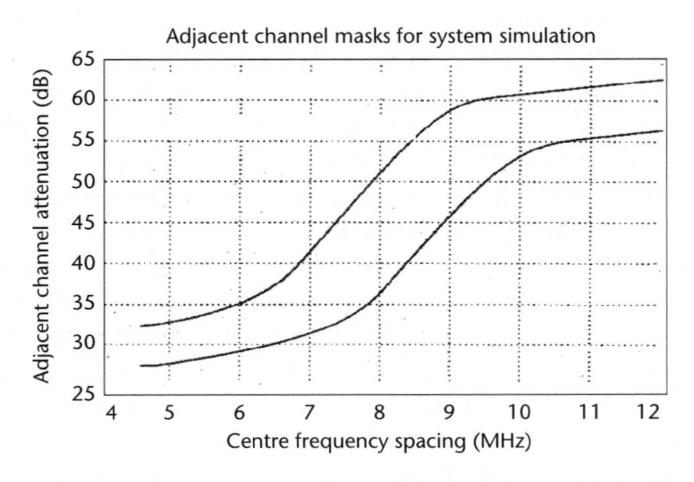

Figure 5.4 ACI (based on ACI masks: WCDMA: Towards IP Mobility and Mobile Internet, *Ojanpera and Prasad).*

reduced, by attenuating the adjacent channel. Link level performance degradation will determine how a non-linear power amplifier will perform with regard to spectrum efficiency. If a user is experiencing interference and at the same time also experiences a reduction in power, which is received at the adjacent channel, this is classed as adjacent channel attenuation.

The ACI simulations can be performed with offset quadrature phase shift keying (OQPSK) chip modulation (OQPSK and quadrature phase shift keying (QPSK) are explained further in this chapter). For this example, two separate ACI masks were created, thus enabling system simulations to be performed. Use of a frequency mask provides over-protection particularly at frequency offsets away from the corresponding adjacent channel. Plots of these masks can be seen in Figure 5.4. It is possible to produce such masks by forcing a power reduction of 3.5 dB within the power amplifier, and by using a receiver 'square root raised cosine filter' with the identical roll-off factor found in the first amplifier, the actual spectrum spreading power can be measured.

These same ACI masks can also be used for downlink simulations, for both OQPSK and QPSK modulation; the measured amplifier effects are quite similar to each other. To briefly define these two modulation schemes, OQPSK modulation is a scheme whereby a delay is introduced being equal to one half the duration of a chip being modulated. This can be best understood by referring to Figure 5.5. Introducing this delay (as can be seen in the OQPSK diagram in Figure 5.5) into one of the quadrature branches of the processing elements, enables the phase to change in both processing elements simultaneously. Hence, the phase shift only occurs in steps of ±90°, thus ensuring that a 180° shift is never required as with QPSK. In the case of 180° phase shifts, the signal will pass through a 180° phase transition and thus the signal envelope will momentarily collapse and reach zero. This zero crossing requires a vast amount of dynamic range from the power amplifier, and limiting these transitions to 90° will avoid amplifier linearity problems which are caused when the input is varied with respect to its amplitude. As can be seen in Figure 5.5 the QPSK modulation scheme is almost identical to the OQPSK scheme, the only difference being that QPSK does not have a delay added into the quadrature branch. Hence, with QPSK large phase shifts of 180° will occur. In order to facilitate these large phase shifts within QPSK, a high quality linear power amplifier will be required which can be installed

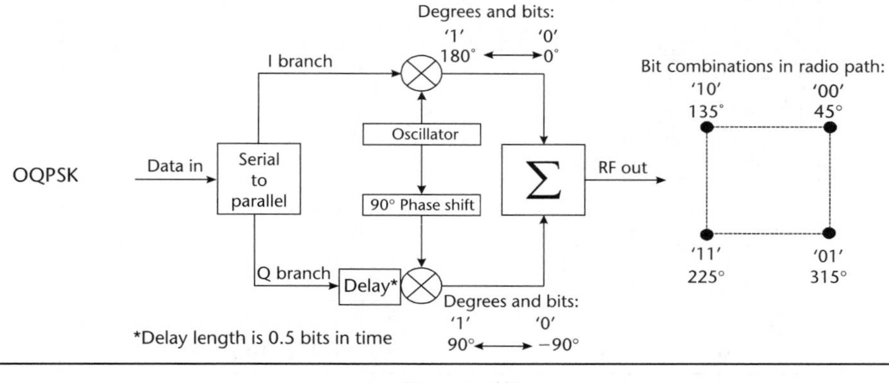

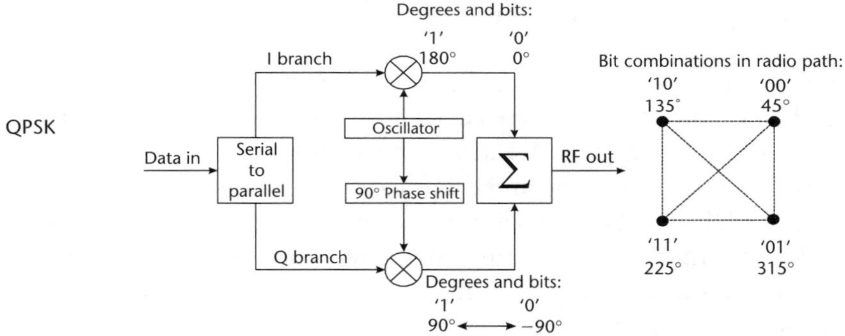

Figure 5.5 Modulation schemes.

in the base station. It is not practical to incorporate such a linear power amplifier within the UE due to size and power constraints. This is why these two different modulation schemes are used: OQPSK for the uplink, where we are constrained by peak to average power ratios, and QPSK for the downlink where we are not so constrained. In summary, by using this system it will be possible to bypass high data rates as required in both the uplink and the downlink. In addition, if the adjacent channel attenuation can be maintained at a sufficient level to reduce the ACI components, it will be possible to maintain the required coverage.

5.5.4 An Adjacent Channel Interference Scenario

To put ACI into context the following scenario illustrates one of the most likely problems due to occur and its effects on coverage. The basis of this scenario is the problems associated in a situation when a UE is executing

Parameters	Values
UE transmit power	21 dBm (0.125 W)
Microcell noise figure	6 dB
Minimum coupling loss – UE and microcell	55 dB
ACIR	33 dB
Thermal noise level (W-CDMA)	−110 dB
Thermal noise level in microcell base station receiver	−110 dB + 6 dB = −104 dB
ACI	21 dBm − 55 dB − 33 dB = 67 dB
Noise rise caused by ACI	−67 dB + 110 dB = 43 dB

Table 5.4 Uplink adjacent channel interference (ACI) – worst case scenario.

maximum transmission power and is located next to a base station currently receiving on a separate adjacent carrier. The conditions for this scenario are shown in Table 5.4 in which a coupling loss of 55 dB has been assumed. The coupling loss can be defined as the minimum pathloss between the UE and the base station antenna. In such a situation it is inevitable that the high interference signal level increase will disrupt the microcells uplink coverage area. The ACI received in the microcell base station is 43 dB greater than the thermal noise level in the receiver. As depicted in Table 5.4, we can see that the ACI in the microcell base station receiver is −67 dBm. This may not occur in all locations as it would require operator X's UE to be transmitting at full power (21 dBm), while simultaneously operator X's UE would have to be located very near to operator Y's microcell antenna. Assuming the UE is located next to the base station antenna, the minimum coupling loss of 55 dB would occur. In addition for this situation to exist, it would require operator X's UE to be transmitting on an adjacent carrier.

In summary, it is important that the planner is fully aware of the various interference issues discussed in this section, as keeping all ACI to a minimum will result in increased coverage.

5.6 Uplink Simulations

It is possible to perform simulations using software-driven applications that are designed specifically for this purpose. One well-known simulation tool is the communications simulation and system analysis program (COSSAP). COSSAP is covered in more detail in Section 2.3.6. This is known as a stream-driven type simulation program. Stream-driven simulations provide algorithmic system-level designers with simulation capabilities at the concept, algorithm and architecture levels. A stream-driven simulator is based on a full dynamic data flow approach, thus enabling activation as soon as all the data elements needed to perform the processing are available. This is advantageous as it is possible to process data at more than one rate compared with conventional digital signalling processing applications and hence this gives incomparable performance as against conventional digital signalling processing.

The 'C' programming language is the dominant language used to construct this type of simulation program, as all the receiver blocks can be generated both in a flexible and efficient manner. A system simulator can be used to study the effects of ACI from a UE operating in a macrocell environment, to the microcell environment. Such simulations can be performed using various specific simulation programs. A scenario likely to occur will be where a UE is located, say, within a macrocell, but is maintaining an actual connection to a microcell which could be located within the boundary of the macrocell. This is likely to occur as the UE will in reality be closer to the base station of the microcell, hence the pilot channel received from the microcell (explained in further detail in Section 3.6) will be stronger than the pilot channel from the macrocell. Although the scenario discussed in Section 5.5 shows high ACI levels, it is nevertheless beneficial to attempt an ACI analysis simulation, thereby providing results for additional ACI scenarios. For this uplink simulation both macro- and microcells were included. A somewhat typical value estimated to be 54 dB for this simulation was assumed for the coupling loss (see Table 5.5). No overhead load has been assumed for the uplink and all assumed connections are for 8 kbps speech.

Simulation parameters	Values
UE maximum power	21 dBm
Noise rise in macrocell (maximum)	6 dB
Noise rise in microcell (maximum)	20 dB
Distance between macrocell base station	1 km
Distance between microcell base station	190 m
Minimum coupling loss	54 dB

Table 5.5 Uplink simulations (based on WCDMA for UMTS, *Holma and Toskola, Wiley).*

For simulation purposes in the downlink, it has been assumed that the signalling overheads were created in a manner whereby the base station's transmission power was increased to compensate for the control channels.

Noise rise must also be taken into consideration and can be defined as the increase in wideband interference level over the thermal noise received in the base station.

Within macrocells the maximum permitted noise rise in the uplink was 6 dB, whereas 20 dB is the level in the microcells. Noise rise is quite significant within macrocells, as a higher noise rise will cause the coverage area to be reduced; a minimal rise is however permitted. The difference in noise rise is considerably lower for macrocells due to intracell interference. In addition, interference generated from saturated neighbouring microcells will also cause a reduction in the macrocell coverage area. The maximum cell loading must also be taken into consideration, and hence a pre-specified maximum noise rise value has to be selected. This value should be maintained regardless of any ACI, thus preventing any further reduction of the cell range.

Real-time radio resource management algorithms that will ensure the uplink loading is not exceeded are performed by the admission control algorithms and the packet scheduling and load control. These are covered in more detail in Chapter 8, entitled 'Radio Resource Management'. Due to the ACI causing capacity reductions both the macro- and microcell coverage areas were kept within the specified limits for the simulations. Capacity is not so responsive to ACI, as can be seen in Figure 5.6 when an ACIR figure of higher than 20 dB exists between carriers.

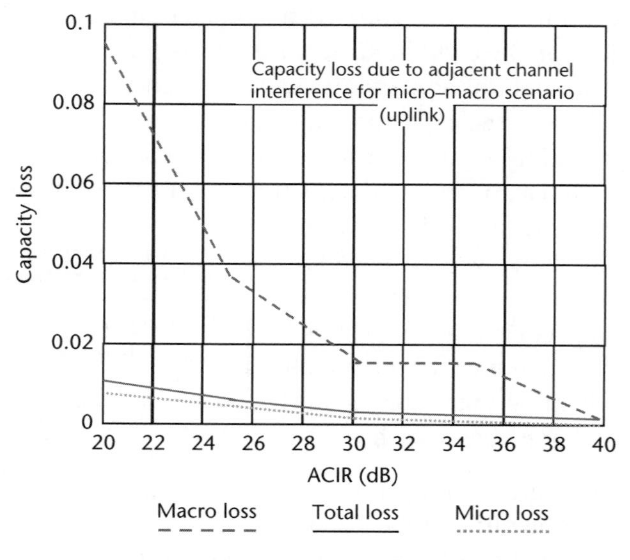

Figure 5.6 Uplink capacity loss due to ACI (reproduced by permission of Wiley & Sons: WCDMA for UMTS, *Holma and Toskola).*

Macrocell capacity suffers more from ACI than microcells because there are fewer interfering users within a macrocell environment, so consequently less interference is generated. By contrast, microcells are generally located in more densely populated areas, thus an increased number of users will generate a significant amount of interference. This has the effect of causing users within any adjacent macrocells also to experience interference.

5.6.1 Planning Adjacent Channel Interference

The ACI can be significantly minimized if a high coupling loss exists between the UE and the base station. Antenna positioning should be taken into consideration, as this will effect the RF propagation. Desensitization can prove beneficial in reducing the ACI as it will reduce the base station sensitivity, in turn increasing the noise figure of the receiver's front-end. Hence, the base station will be less sensitive to any ACI. This will prove disadvantageous as it will cause a reduction in the cell size, as the base station receiver is unable to detect weak signals from UEs located at the cell's periphery. This method is advantageous when dealing with small cells, as uplink coverage is not considered to be a problem in such an environment. The ideal or optimum solution to reduce ACI is to use different frequency bands as, when dealing with two separate frequencies, there will be less chance of ACI occurring. If both sets of antennas can be located on the same masts, similar power output levels can be set for both base stations and hence, the received signals will be the same in both operators' UEs. This will prove advantageous as, if the base station output powers remain at the same level, the possibility of any ACI spill over will be minimized in contrast to the unwanted carrier being transmitted at a higher power level.

In summary, the planner needs to take into account all possible interference that may be generated, as described in this section, from other operators to any ACI that may be generated via another RF carrier.

In addition, performing uplink simulations can assist in providing a clearer picture with regard to ACI once the cell plan is known and the necessary parameters previously discussed can be substituted into a software simulation package. As it is somewhat difficult to predict how interference affects both micro- and macrocells, simulations can give a useful insight into how the network would behave.

5.7 UMTS Radio Carriers

The RF carriers that have been allocated for UMTS use can be seen in Figures 5.7 and 5.8. As can be seen, the carrier spacings between each carrier are somewhat limited, hence this needs to be carefully managed to

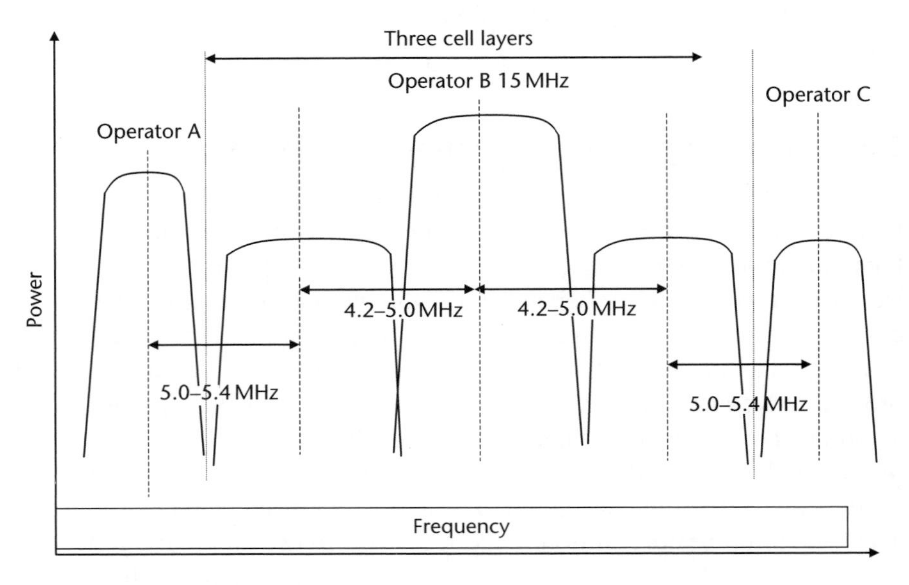

Figure 5.7 UMTS frequency utilization.

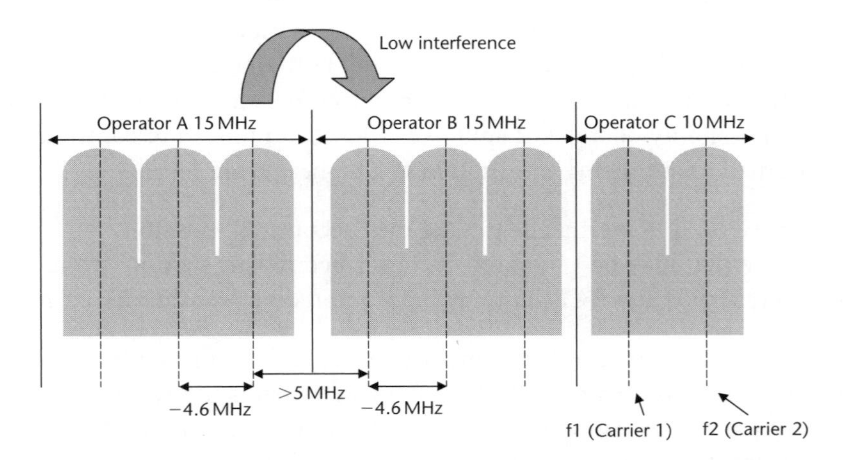

Figure 5.8 UMTS carrier spacings within each operator's band and between separate operators (based on WCDMA for UMTS, *Holma and Toskola).*

ensure that any ACI that may be generated can be minimized as much as possible. This is due to the fact that the available frequency spectrum allocated for UMTS is limited.

If one carrier is located too close to the next carrier ACI will occur. However it is possible to adjust the carrier spacing on increments of a 200 kHz raster, which would therefore result in a reduction of ACI. If this spacing were to be increased, then further ACI reductions could be achieved, nevertheless as the spectrum allocated is limited, only minimal spacings can be achieved for both operator and inter-operator scenarios. If the same base station antennas utilize both carriers, then the ACI would be minimal, so in this particular scenario it would be possible to reduce the carrier spacing to around 4.0 MHz, thus freeing up more of the valuable spectrum. As illustrated in Figures 5.7 and 5.8, carriers can be shifted in steps of 200 kHz that vary, for example, from 4.2 to 5.4 MHz. In this particular case, the operator in question possesses three continuous blocks of the spectrum (15 MHz), however this may not be the situation in all networks as the block allocation may be separated by other operators. A basic indication of carrier spacings is illustrated in Figure 5.8. Spacing can be thought of as locating the W-CDMA carriers within the available frequency spectrum, but with sufficient space between each carrier, therefore ensuring that good isolation can be achieved. In simple terms as illustrated in Figure 5.8, the separation between carriers A and B is greater than between carriers B and C, hence less interference will be generated between the former carriers A and B. Isolation is necessary as it will reduce any intermodulation or interference issues which would occur if sufficient spacing between the carriers did not exist. Ultimately the greater the carrier spacing, the greater the protection against potential interference from any adjacent unwanted carriers.

5.7.1 Carrier to Interference Ratio

The *C/I* ratio can be determined as the ratio of the carrier energy (*C*) to the interference energy (*I*). The illustration in Table 5.6 shows some examples of the *C/I* ratio given in decibels. For example, to achieve a required *C/I* ratio of 10 dB, then the *C* must have a 10 dB higher power level than the interfering signal (*I*). A variety of parameters will influence the optimal

Table 5.6 C/I ratio.

C/I (dB)	*I* (dBm)	C (dBm)
20	−100	−80
−18	−118	−100
10	−100	−90
−10	−80	−90*

* Interference signal is 10dB stronger than the carrier.

C/I level required. Tests can be performed to evaluate if the required *C/I* ratio could be increased or reduced in relation to any equipment vendor values that may be specified. Alternatively, the actual equipment design specifications can be analysed.

5.7.2 Guard Bands and Zones

As previously discussed earlier in this section, the bandwidth to be used for UMTS will be 5 MHz (see Figure 5.9). The 10 and 20 MHz alternatives

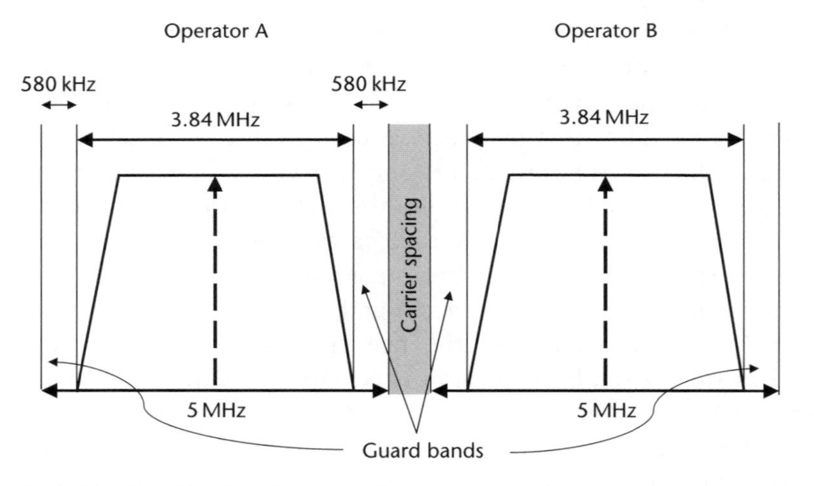

Figure 5.9 Guard bands and zones.

will offer greater capacity, however using the entire frequency band may not be viable in certain countries, and therefore the occupancies within the allocated frequency band may set certain limits. As can be seen in Figure 5.9, the usable bandwidth within W-CDMA is 3.84 MHz, however with the required guard bands the total bandwidth amounts to 5 MHz. Guard bands can be best thought of as a zone around the W-CDMA coverage area which will assist in preventing any unwanted co-channel interference. The guard bands will also assist in providing noise floor rejection, by preventing the use of that portion of the carrier's frequency roll-off. The roll-off can be thought of as the leading and falling edge of the carrier. The noise floor can be considered as the level of noise that is present in the carrier, which generally refers to any unwanted signal. The guard bands also protect against interference generated from other systems. Consequently GSM systems must be considered as an interferer to a UMTS system.

With this in mind, assuming the ratio between the cell sizes and distances is known, the minimum separation between GSM and W-CDMA base stations using the same frequency can be calculated. To ensure the system does not suffer any blocking by the adjacent spectrum and also to ensure any other RF transmissions do not block the W-CDMA system, adequate guard bands must be used. It is therefore possible for other radio systems on the same band to create co-channel interference. This situation has already transpired in North America within the personal communications systems (PCS) arena. For example, in the USA there are two bodies existing which deal with these types of issues. The National Spectrum Managers Association (NSMA) is responsible for dealing with issues associated with the inter-system co-ordination specifications and interference calculation parameters. In addition, the Code of Federal Regulations (CFRs) is responsible for dealing with inter-system interference issues.

To conclude, the use of guard bands should always be considered as they will assist in reducing interference levels, which in turn will assist in performance maintaining the CCQ.

5.7.3 Frequency Sharing

Frequency sharing will enable operators within a specified geographical region to share the entire allocated frequency spectrum, which has political, commercial and operational advantages. It will allow more operators to enter the 3G market and offer UMTS services. As site sharing is likely to be more prevalent with UMTS (3G) implementation due to political and practical restraints, frequency sharing will be possible as W-CDMA systems are planned to have a dynamic channel allocation (DCA) scheme. In addition, with frequency sharing, a reduction in trunking losses can be achieved. A trunk can be thought of as a transmission channel between two points.

Therefore, the planner should be aware that by introducing guard bands around the carrier, performance degradation can be significantly decreased. In addition, GSM systems are likely to cause unwanted interference, so this must also be taken into consideration. Frequency sharing when possible is also advantageous as this can help to resolve unwanted interference, ultimately helping to maintain the coverage required, since, as previously discussed, all unwanted interference will have an adverse effect on coverage.

5.8 Inter-system Handovers (Slotted Mode)

Handovers occur at two levels; one type of handover occurs between 2G and 3G networks and is known as an inter-system handover, alternatively inter-frequency handovers will occur within one operator's system between separate carriers.

To enable inter-frequency handovers, an efficient method is needed to enable measurements to be performed on other frequencies, while still maintaining a connection on the current frequency. There are two methods available for inter-frequency measurements in W-CDMA. The first is slotted or compressed mode, and the second, dual receiver.

5.8.1 Dual Mode

To employ a dual receiver system the UE will require both antenna diversity and a diverse receiver branch. Antenna diversity can provide diversity against interference and fading, in addition to averaging out any receiver noise. This will have an impact on the UE design, its physical size, and costs, as one more receiver would need to be added into the UE. With the dual receiver system, one receiver branch is switched to another frequency to continuously perform the required measurements, while simultaneously the other receiver branch continues receiving on the current operating frequency. Having no physical break in the current frequency connection, gives the possibility to run fast closed loop power control continuously. This is considered advantageous with the dual receiver system, due to the fact that any high speed data connections that are in use will not be disrupted. However, with the continual quest for smaller and cheaper terminals, it is not envisaged at present that the dual receiver system will be widely implemented.

5.8.2 Slotted Mode

The slotted mode approach is more suitable for the UE that does not possess antenna diversity, mainly because it simplifies the UE design and ultimately reduces the costs. The slotted mode system operates by utilizing a space within the W-CDMA frame, which is used to briefly 'listen' to another frequency. During each 10 ms frame the required measurement information is transmitted and both the transmission and reception are momentarily stopped for a few milliseconds to enable alternative frequency measurements to be performed. Taking into account that the default speech codec of UMTS uses 20 ms speech frames, the radio frame to transport these should be at least 20 ms long. Hence, it can be assumed this will be interleaved over 20 ms or two radio frames. A brief illustration of slotted mode is shown in Figure 5.10. The preferred method of slotted

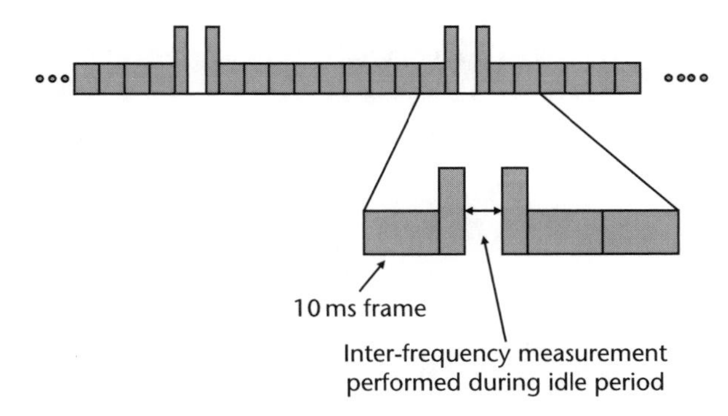

Figure 5.10 Slotted mode.

mode uses a parallel additional scrambling code. In addition, as the processing gain remains the same, the spreading factor is not reduced in the slotted frame. From the illustration in Figure 5.10, it can be seen that a slotted mode system requires that half of one of the 10 ms frames are available for alternative frequency measurement. With this in mind, the information sent must be transmitted in half the normal time, that being 5 ms. It is during this 5 ms period that the UE is able to 'listen' to an alternative frequency, say that of GSM for example.

Finally, inter-system handovers will be necessary within UMTS and, as discussed from the two methods available, slotted mode appears to be the preferred method at this point in time.

In summary, this concludes the chapter on 3G co-planning and co-existence, giving the planner the knowledge required to ensure that co-existence can be implemented with minimal disruption to the networks involved. A good insight has been given into all intermodulation, isolation, interference, and ACL issues, all of which have a detrimental effect on the coverage parameter of the CCQ model. Hence, it is important to be aware of these issues so they can all be efficiently minimized, and, in turn, assist in achieving the maximum coverage possible. Finally, the chapter has shown how performing uplink simulations will provide a good insight into how the network will behave with respect to the required coverage, how and why guard bands are implemented, and the possibilities for inter-system handovers. This completes the section related to coverage issues and leads into the next section, which deals with link budgets and loads.

Influence of Link Budgets on 3G Coverage

6.1 Introduction

This chapter covers link budgets. The text reviews all the parameters required to ensure that both the uplink and downlink budgets can be successfully planned, and the required coverage area achieved.

The link budget identifies and defines the power, size, and location of the cells, while taking into account interference, traffic, and other important variables that could affect the coverage are also covered within this chapter.

Also reviewed in this chapter are energy per bit to the noise spectral density (Eb/No), processing gain, and propagation models. These are followed by some link budget examples, and a study of the relationship of both capacity and coverage to the link budgets.

Finally, because link budgets vary according to the volume of data being transmitted, the planner should always be aware that the radio frequency (RF) network plan will also need to be optimized with regard to the wide range of varying data speeds of the subscribers present within the network. This is one of the most important factors in maintaining the required network coverage. These issues are covered here, and discussed further in the optimization section – Part V.

6.2 Link Budgets and Loads

The three major issues that should be taken into account when planning link budgets are the interference margin, the fast-fading margin, and the soft handover gain. In addition to these, planners should understand the effects of Eb/No and the processing gain.

The interference margin will have an effect on the cell size; the fast-fading margin can be compensated for by power control mechanisms and the soft handover gain can actually give some gain back into the link budget. Finally, once both the uplink and downlink budgets have been calculated, then the results can be input into the propagation model which will then provide the predicted coverage for the cell/sector in question.

6.2.1 Interference Margin

The interference margin is related to the cell loading; when one increases the other also increases. The limiting factor is that, as the interference margin increases, the coverage area will be reduced. However, when considering coverage limited cases a larger interference margin should be used to enable the maximum permissible coverage. The size of the cell is limited by the maximum allowed pathloss in the link budget (with respect to coverage limited cases). In a scenario where a 20–50 per cent cell load is present, then typical values in a coverage limited case for the interference margin will be around 1–3 dB.

6.2.2 Fast Fading

Fast fading, also known as Rayleigh fading, can be thought of as multi-path fading where there is not a clear path between the transmitter and the receiver (TRX). This type of fading is prevalent and occurs when the receiver or the user equipment (UE) are moving, and is due to irregular signal strength variations occurring at the receiving antenna. The end result of this even when the signal paths are added together is an overall reduction in the signal strength and is therefore disadvantageous. Macro diversity gain can help to reduce fast fading when the UE is in a soft handover state, when the desired signals are received by more than one base station and are summed together at the radio network controller (RNC) (referred to as macro diversity) ensuring that the wanted signal can be constructed from several sources, and strengthening the reduced signal levels caused by fast fading. Alternatively slow fading, also known as shadowing, occurs when there is a partial blockage between the TRX and can be caused by such obstructions as buildings and foliage. This 'clutter' can cause partial absorption of the signal and therefore the signal strength is again reduced.

When taking closed-loop power control into account, there will be some headroom required within the transmission power level of the UE. This becomes more apparent with UEs that are moving slowly whereby fast fading can be compensated for by the closed-loop power control. For slow-moving UEs, typical values for fast fading consist of around 2 and 5 dB.

6.2.3 Soft Handover Gain

If the required log-normal fade margin is reduced (log-normal fading), then both soft and hard handovers will offer a gain against slow fading. This is due to the fact that slow fading is partly uncorrelated between base stations and the UE is able to select a more appropriate base station by performing a handover. Soft handovers are able to reduce the Eb/No due to the additional macro diversity gain against fast fading. The reduction in Eb/No, relative to a single radio link, can also be achieved due to the effect of macro diversity combining. Macro diversity, as described above, exists at the RNC level and when a UE is in a soft handover scenario, the required signals are received by more than one base station. In these cases the base stations perform the signal summing first with regard to their own radio paths, but the final signal summing is performed at the RNC. This enhances the original signal strength as it is summed together from several sources.

The total soft handover gain, including the gain against both slow and fast fading, is assumed to be between 2 and 3 dB.

6.2.4 Energy per Bit per Noise Spectral Density

The signal-to-noise ratio defines a receiver's performance while experiencing unwanted noise within a wideband-code division multiple access (W-CDMA) system and this can be equated to the Eb/No.

A standard value within most CDMA systems is about 7.5 dB of power over the effective noise floor and this can then be added into a link budget calculation (covered later in this chapter). The value of the Eb/No has a distinct relationship with the TRX capacity and must be considered when attempting to plan the link budget. The Eb/No value will vary depending on the speed of the UE. If the UE is moving at a slow speed then the Eb/No requirement will be relatively low, in the region of around 1.5 dB. However, the fast-fading margin must also be taken into account, which for a slow-moving UE would be approximately 4 dB and so in this particular case the coverage dimensioning will be limited due to the low speed of UE.

An example of how to calculate the TRX capacity is as follows:
(*This example assumes that the total number of users connected to one TRX required to transfer 60 kbps baseband rates and the Eb/No relationship is 1.6 dB.*)

$$\text{TRX capacity, } X \approx \frac{Gp}{Eb/No} = \frac{(38,40,000/60,000)}{1.6} = \frac{64}{1.6} \approx \underline{\underline{40 \text{ users}}} \qquad (6.1)$$

where Gp is the processing gain (384 megachips per second (Mcps), refer to Section 6.2.4.1, Processing Gain) and Eb/No, the energy per bit/noise spectral density.

Note: This is one method that can be utilized to estimate TRX capacity. Other issues, which are covered later in this chapter, must also be taken into account. However, in this context, the number of users can be equated to the number of channels, thus identifying how many channels/users may use the TRX with the existing signal-to-noise ratio level.

6.2.4.1 Processing Gain

The ratio of the transmitted bandwidth to the information bandwidth is called the processing gain (Gp). The transmission bandwidth (Bt) must always be larger than the information bandwidth (Bi):

$$\text{Processing gain, Gp} = \frac{\text{Bt}}{\text{Bi}} \qquad (6.2)$$

In addition, the bandwidth is statistically independent of the information-bearing signal, therefore the resulting RF bandwidth is determined by a different function than the actual data being sent. When high data rate usage occurs, the processing gain will be low and hence the coverage area will be reduced. Conversely, with low data rate usage the processing gain will be high and therefore the coverage area will be increased.

6.2.4.2 Eb/No Requirements as a Function of Load

As an increase in load occurs beyond 50 per cent, then the maximum cell range decreases significantly. As can be seen in Figure 6.1, the Eb/No

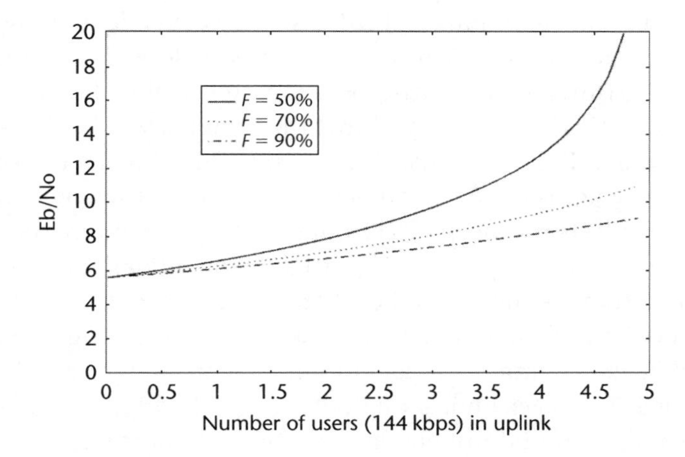

Figure 6.1 Eb/No requirements as a function of load in different propagation environments. Source: WCDMA Towards IP Mobility and Mobile Internet, *Ojanpera and Prasad.*

requirements as a function of number of users is shown for the fractional cell loading (*F*), equalling 50, 70, and 90 per cent, with regard to the different propagation environments. Fractional cell loading can be considered as a partial number of users present within a W-CDMA network.

The propagation environment will have a marked effect on the value of the fractional cell loading and hence the greater the pathloss attenuation factor, the higher the value of '*F*'. When '*F*' is equal to 50 per cent, the Eb/No requirements are much higher than when '*F*' is equal to 70 per cent.

This occurs as there is high intercell interference, as well as an increase in intracell interference. With an improved Eb/No performance in the base station it is possible to enhance both the system capacity and the coverage. This results from the Eb/No requirement being present in both the load equation and the link budget.

6.2.5 Uplink Budget

The illustration in Figure 6.2(a) shows a block diagram of how an uplink budget should be constructed. All losses and gains from the UE side must be taken into account including body loss into the link budget and the effect of attenuation caused by the human head. In general it is dependent on the operational conditions and will also vary dependent on the environment and the user. Typical values for body loss are considered to be about 3 dB. However, in third-generation universal mobile telephony system (UMTS (3G)) it is expected that the terminal will be operated in a different manner. Due to the visual front-end of the 3G handsets compared to the mainly voice service second-generation (2G) terminals, it is more likely that handsets will be held in front of the user therefore there will be reduced signal losses previously caused by the influence of the user's body. Penetration loss must also be taken into consideration, as this will affect the operational environment. Actual values can vary significantly between different environments and ideally local measurements should be performed. Typical examples would be around 15 dB for an indoor environment and around 8 dB for an in-vehicle environment. The antenna gain accounts for the gain of the UE antenna and typical values are around 0 dBi (dBi refers to a decibel, that is a unit of power referenced to the gain of a theoretical isotropic radiator) although it is likely that some data terminals will have antennas giving a marginally higher gain of around 2 dBi.

6.2.6 Downlink Budget

To simplify this, take the maximum uplink pathloss shown in Figure 6.2(b) that will equal the totals of losses from both the UE and the base station (which have been derived from Figure 6.2(a)), and 'plug' these combined losses in to the downlink budget (i.e. the maximum uplink acceptable pathloss as shown on the left-hand side of Figure 6.2(b)). The next step is to add in the reliability margin and environment correction to obtain the

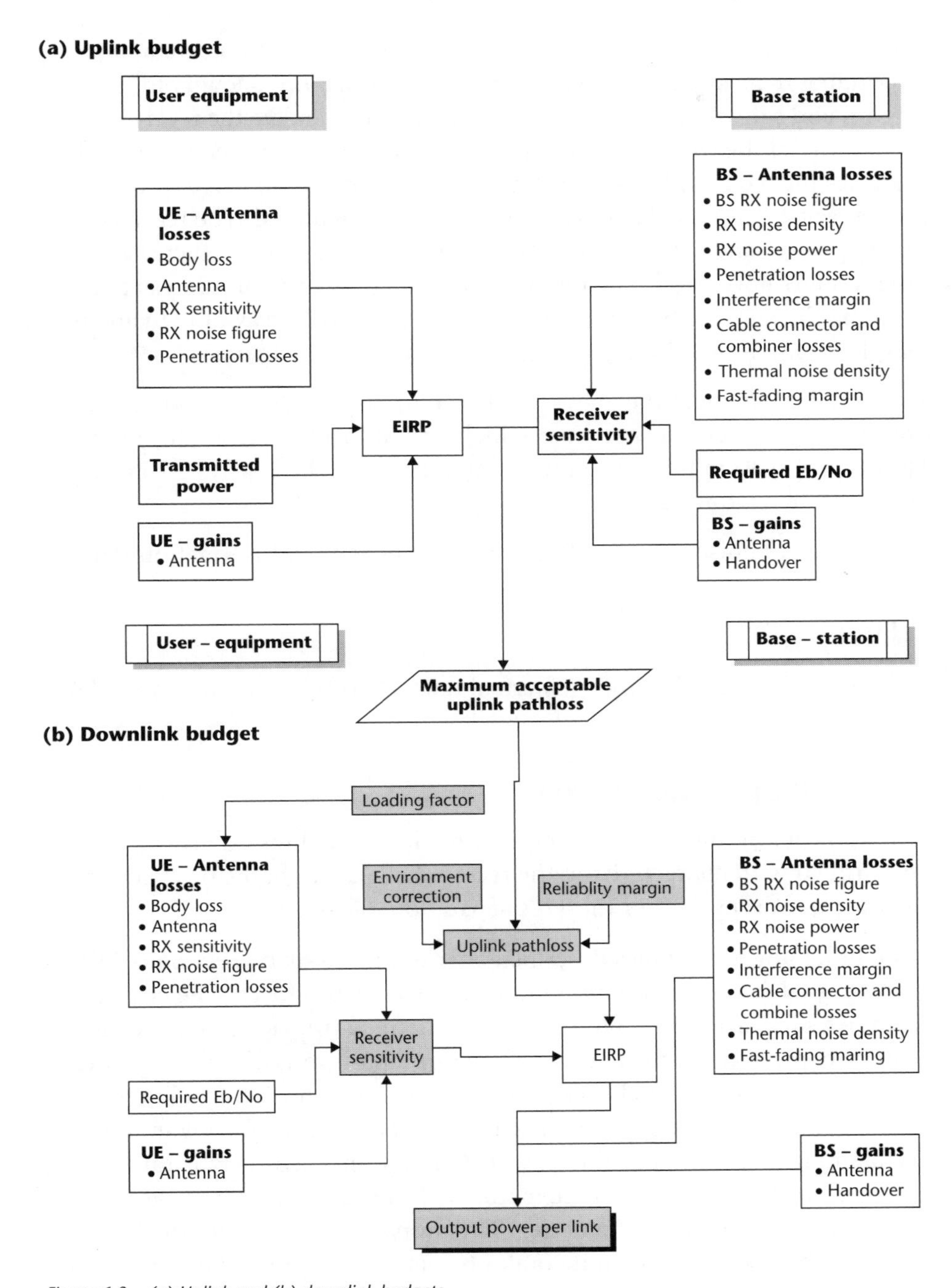

Figure 6.2 (a) Uplink and (b) downlink budgets.

downlink power requirement. Keeping in mind the downlink element, by adding the losses and the gains from the base station (shown on the right-hand side of Figure 6.2(a)), it is possible to calculate the receiver sensitivity of the UE. This is the signal level needed at the receiver input that satisfies the required Eb/No. An activity factor must be added, to provide more realistic results, if real receiver sensitivity is required. The receiver noise power can be defined mostly as thermal noise, which is increased by the receiver noise level. The thermal noise can only be reduced by actually cooling the receiver, but this is not feasible. So the only way of reducing the level of noise is by selecting a receiver with a low-noise figure.

The final outcome of the above, as shown in Figure 6.2(a and b), is the output power from the antennas which will have an effect on the amount of RF propagation from the antenna and therefore the level of coverage over the area required:

$$\text{Receiver noise power} = \text{Receiver noise density} + \text{Information rate} \tag{6.3}$$

Thus:

$$\text{Receiver noise density} = \text{Thermal noise density} + \text{Receiver noise level}$$

6.2.7 Propagation Model

An RF propagation model will determine the RF pathloss. This is performed with characterizing the received signal strengths by averaging out the power levels over the TRX separation distances.

The link budget examples described later in this chapter (Section 6.2.9) can be used in the propagation model to provide distances. The coverage area per site (measured in km^2/site) for a defined propagation environment can be expressed as the coverage efficiency. Apart from the coverage area the traffic density must also be taken into account and, using a known propagation model (e.g. the Walfisch–Ikegami model or the Okamura–Hata model), the propagation loss calculated from the previous link budget can be inputted into the model and thus converted into the maximum cell range in kilometres. The propagation loss from the first two link budgets (as shown in Tables 6.3 and 6.4 found and described in Section 6.2.9) were inputted into the Okamura–Hata model. This example is for an urban macrocell taking into account an average UE height of 1½ m and a base station antenna height of 30 m. For the suburban area, an additional correction factor of 8 dB has been assumed as this provides a more consistent result pattern. The calculated cell ranges are shown in Table 6.1 and will be explained in more detail in Section 6.2.9, which deals with the three link budget examples comparing different data rate usage.

Variables	Propagation model examples		
	Example	Pathloss	Cell range (km)
UE height = 1.5 m Antenna height = 30 m	1: 12.2 kbps speech service	141.9	2.3
Area correction factor = 8 dB	2: 144 kbps RT data	133.8	1.4

Table 6.1 Propagation model.

6.2.8 Sample Budget Calculation

A link budget can be thought of as the basis for network dimensioning (as reviewed in Chapter 4) and it is possible from the calculation of a link budget to ascertain an initial indication of the cell range. It will then be possible to calculate the number of sites required within the specified coverage area. A link budget can prove useful in identifying the parameters requiring improvement in order to enhance the coverage area (this is discussed in more depth in Part V, entitled Optimization and Network Planning). The same principles for global system for mobile communications (GSM) link budgets are applied for the UMTS (3G) link budgets, with the only difference being that more parameters are considered; for example Eb/No requirements, processing gain and soft handover gain, etc. Table 6.2 (*Note:* This is a sample table to be used for instructional purposes only.) shows one sample of a link budget calculation, indicating the total pathlosses in the uplink and the downlink.

6.2.9 Link Budget Examples

In the three tables found in this section (Tables 6.3–6.5), typical link budget examples are provided for 12.2 kbps (using an adaptive multi-rate (AMR)

Table 6.2 Sample link budget calculation template.

	Parameter	Downlink	Uplink	Unit
(a)	Average TX power/TCH	30	24	dBm
(b)	Cable, connector, and combiner	2	0	dB
(c)	Transmitter antenna gain	13	0	dBi
(d)	TX EIRP/TCH = a − b + c	41.0	24.0	dBm
(e)	Receiver antenna gain	0	13	dB
(f)	Cable, connector, and splitter losses at the receiver	0	2	dB
(g)	Receive noise figure	5	5	dB
(h)	Thermal noise density	−174	−174	dBm/Hz
(i)	Receiver interference density	−1000	−1000	dBm/Hz
(j)	Total noise + interference density (g + h + i)	−169	−169	dBm
	Information rate (R_b)	8	8	kHz
(k)	10 log(R_b)	39.03	39.03	dBHz
(l)	Eb/(No + Io) (link level similar result)	8	6.6	dB
(m)	Receiver sensitivity (j + j + l)	−122.0	−123.4	dBm
(n)	Handoff gain	5	5	dB
(o)	Other gains	0	0	dB
(p)	Log-normal fade margin	11.3	11.3	dB
(q)	Maximum pathloss (d − m + (e − f) + n + o)	153.67	149.07	dB
(r)	Range	4.84	3.61	km

Source: WCDMA: Towards IP Mobility and Mobile Internet, Ojanpera and Prasad.

speech codec), 144 kbps real-time (RT) data and 384 kbps non-real-time (NRT) data services, all of which take into account the planned uplink noise rise in an urban macrocellular environment. The following three sample link budgets are calculated taking into consideration the assumed TRX parameters shown in Tables 6.6 and 6.7.

6.2.9.1 Link Budget Example – 12.2 kbps

The sample link budget illustrated in Table 6.3 is for an AMR 12.2 kbps voice service for in-vehicle users at speeds of 120 km/h, which allow for soft handovers and also include an 8.0 dB loss for in-car use. The fast power control, at physical speeds of 120 km/h, is unable to compensate for the fading, and therefore no fast-fading margin is reserved.

Table 6.3 Link budget example – 12.2 kbps.

12.2 kbps voice (vehicular 120 km/h)		
UE		
Maximum UE TX power (W)	0.125	
Maximum UE TX power (dBm)	21.5	a
UE antenna gain (dBi)	0.0	b
Body loss (dB)	3.0	c
EIRP (dBm)	18.0	d = a + b + c
Base Station		
Thermal Noise Density (dBm/Hz)	−174.0	e
BS receiver noise figure (dB)	5.0	f
RX noise density (dBm/Hz)	−169	g = e + f
RX noise power	−103.2	h
Interference margin	3.0	i
RX interference power	−103.2	j
Total effective noise + Interference (dBm)	−100.2	k
PG (dB)	25	l
Required Eb/No (dB)	5.0	m
Receiver sensitivity (dBm)	**−120.2**	**n = m − l + k**
BS antenna gain (dBi)	18.0	o
Cable loss in BS (dB)	2.0	p
Fast-fading margin (dB)	4.0	q
Maximum pathloss (dB)	**154.2**	**r = d − n + o − p − q**
Coverage probability (%)	95	
Log-normal fading constant (dB)	7.0	
Propagation model exponent	3.52	
Log-normal fading margin (dB)	7.3	s
Soft handover gain (dB), multi-cell	3.0	t
In-car loss (dB)	8.0	u
Allowed propagation loss for cell range (dB)	**141.9**	**v = r − s + t − u**

Source: WCDMA for UMTS, Holma and Toskola.

The required Eb/No is assumed to be 5.0 dB and this is dependent on:

a) data rate,

b) multi-path profile,

c) UE speed,

d) type of service (TOS),

Table 6.4 Link budget example – 144 kbps.

144 kbps RT data (indoor user covered by outdoor BS – 3 km/h)		
UE		
Maximum UE TX power (W)	0.25	
Maximum UE TX power (dBm)	24	a
UE antenna gain (dBi)	0.0	b
Body loss (dB)	0	c
EIRP (dBm)	26.0	d = a + b + c
Base Station		
Thermal noise density (dBm/Hz)	−174.0	e
BS receiver noise figure (dB)	5.0	f
RX noise density (dBm/Hz)	−169	g = e + f
RX noise power	−103.2	h
Interference Margin	3.0	i
RX interference power	−103.2	j
Total effective noise + Interference (dBm)	−100.2	k
PG (dB)	14.3	l
Required Eb/No (dB)	1.5	m
Receiver sensitivity (dBm)	–113	n = m − l + k
BS antenna gain (dBi)	18.0	o
Cable loss in BS (dB)	2.0	p
Fast-fading margin (dB)	4.0	q
Maximum pathloss (dB)	151	r = d − n + o − p − q
Coverage probability (%)	80	
Log-normal fading constant (dB)	12	
Propagation model exponent	3.52	
Log-normal fading margin (dB)	4.2	s
Soft handover gain (dB), multi-cell	2.0	t
In-car loss (dB)	15.0	u
Allowed propagation loss for cell range (dB)	**133.8**	**v = r − s + t − u**

Source: WCDMA for UMTS, Holma and Toskola.

e) receiver algorithms,

f) base station antenna configuration.

For UEs travelling at low speeds, the Eb/No requirement is low, but a fast-fading margin will be required. The fast-fading margin causes low UE

Table 6.5 Link budget example – 384 kbps.

384 kbps RT data (outdoor user 3 km/h – no soft handover)		
UE		
Maximum UE TX power (W)	0.25	
Maximum UE TX power (dBm)	24	a
UE antenna gain (dBi)	0.0	b
Body loss (dB)	0	c
EIRP (dBm)	26.0	d = a + b + c
Base station		
Thermal noise density (dBm/Hz)	−174.0	e
BS receiver noise figure (dB)	5.0	f
RX noise density (dBm/Hz)	−169	g = e + f
RX noise power	−103.2	h
Interference margin	3.0	i
RX interference power	−103.2	j
Total effective noise + Interference (dBm)	−100.2	k
PG (dB)	10	l
Required Eb/No (dB)	1	m
Receiver sensitivity (dBm)	**−109.2**	**n = m − l + k**
BS antenna gain (dBi)	18.0	o
Cable loss in BS (dB)	2.0	p
Fast-fading margin (dB)	4.0	q
Maximum pathloss (dB)	**147.2**	**r = d − n + o − p − q**
Coverage probability (%)	95	
Log-normal fading constant (dB)	7	
Propagation model exponent	3.52	
Log-normal fading margin (dB)	7.3	s
Soft handover gain (dB), multi-cell	0.0	t
Allowed propagation loss for cell range (dB)	**139.9**	**v = r − s + t**

Source: WCDMA for UMTS, Holma and Toskola.

Assumption	Data terminal	Voice terminal
Antenna gain	2 dBi	0 dBm
Body loss (dB)	0	3
Maximum TX power (dBm)	24	21

Table 6.6 UE assumptions.

Assumption	Parameter
Antenna gain	18 dBi (three-sector base station)
Noise figure	5.0 dB
Eb/No requirement	Voice: 5.0 dB 144 kbps RT data: 1.5 dB 384 kbps NRT data: 1.0 dB
Cable loss induced	2.0 dB

Table 6.7 Base station assumptions.

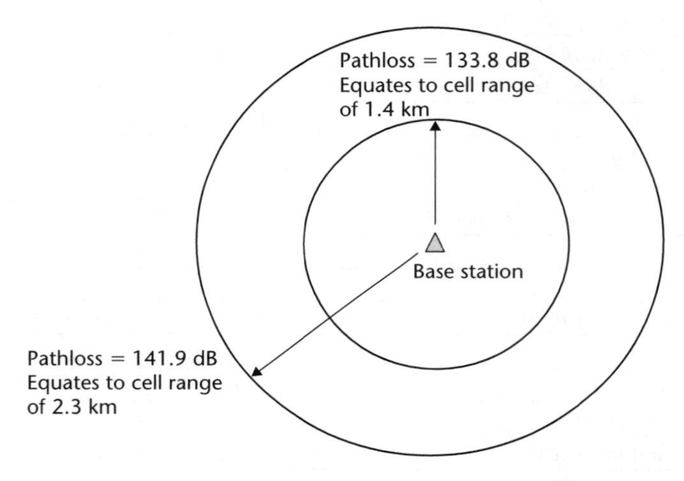

Figure 6.3 Pathloss and cell range.

speeds to be a limiting factor with regard to actual coverage. Figure 6.3 illustrates the correlation between the pathloss and the actual cell range. The actual cell range is calculated from the final pathloss figure using an RF propagation prediction tool, and when converted, there is a direct relationship between the lower pathloss and a reduced cell range.

6.2.9.2 Link Budget Example – 144 kbps

As shown in the last example, slow UE speeds limit coverage. Although the Eb/No requirement is reduced, fast power control must be able to compensate for the fast fading and thus a fast-fading margin headroom must be taken into account. Typical values here are considered to be between 2 and 5 dB. The example illustrated in Table 6.4 shows a link budget for a 144 kbps RT data service, whereby an indoor location probability of 80 per cent is achieved by the outdoor base station with soft handovers. Note the main differences here are:

a) a higher UE transmission power,

b) a lower processing gain,

c) a lower Eb/No requirement.

Additionally, an extra 4 dB are reserved for the fast-fading margin described above, and a typical building penetration loss of 15 dB is assumed.

6.2.9.3 Link Budget Example – 384 kbps

The third example, as shown in Table 6.5, is a link budget (using the UE assumptions shown in Table 6.6) prepared for 384 kbps NRT data in an outdoor environment. This scenario shows that the processing gain is lower than the previous examples as the bit rate is higher and the coverage is diminished due to reduced efficiency in the base station receiver sensitivity. The processing gain is the ratio of transmission bandwidth to information bandwidth and therefore represents the ability of the receiver to disregard unwanted spread signals. It should also be noted that the Eb/No is lower with higher data rates, thus compensating for the reduced coverage of high data rates. The lower the Eb/No requirement means the less power needed for the same performance, and a larger cell radius can be achieved. For this particular example link budget, no soft handovers have been assumed as these will affect the final results as shown in the previous two examples.

For these three link budget examples the final pathloss results will be input into a RF propagation model and the cell range and coverage can then be acquired. This has been explained in further detail in Section 6.2.7.

6.3 Coverage Versus Capacity in Relation to Link Budgets

Coverage is more dependent on load in the downlink than in the uplink, therefore it is important to achieve a balance for both the uplink and the downlink coverage. It is possible to increase capacity by splitting the downlink power output. The comparison between coverage and capacity needs to be taken into consideration when dealing with further gains that can be achieved with the use of transmit diversity.

Coverage is affected by both the uplink and the downlink air-interface load. As can be seen in Figure 6.4 the coverage is more dependent on the

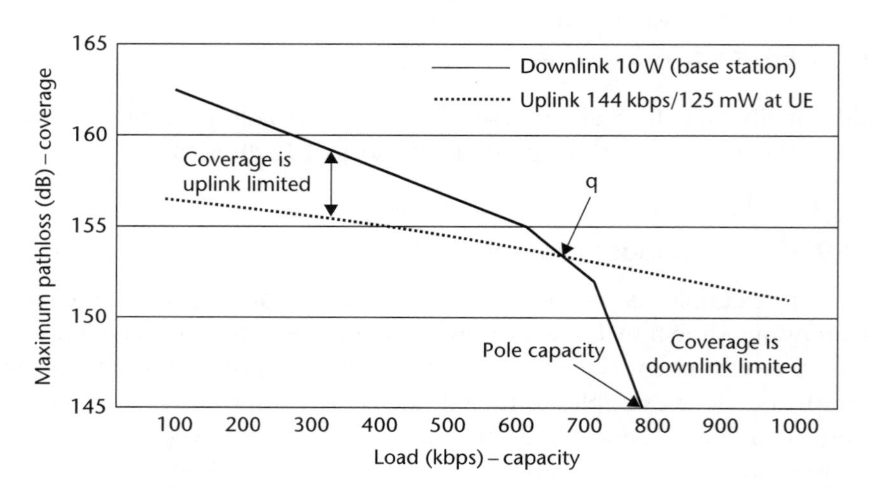

Figure 6.4 Coverage versus capacity (maximum pathloss versus load).

load in the downlink than in the uplink. This can be attributed to the fact that the maximum transmission power from the base station in the downlink in this particular example is 10 W. This power output remains the same regardless of the number of users. If a low traffic load is present in the downlink, as each UE has its own power amplifier, the coverage will be reduced as a function of the number of users present in the cell. This is also shown in Figure 6.4 where the maximum pathloss continues to decrease until the downlink from the base station meets equilibrium with the uplink from the UE at point 'q'. After this intersection of the uplink and downlink, the downlink pathloss drops significantly as the load continues to increase. Figure 6.4 also shows the uplink limited and downlink limited coverage areas affected by this cross-over.

In conclusion, the capacity is limited by the downlink and hence the coverage here is limited by the uplink for a load of 650 kbps. The illustration in Figure 6.4 shows the coverage as a function of load for both the uplink and the downlink, which equates to the maximum pathloss.

Hence a balance must be achieved at the cross-over point to obtain optimum results for both the uplink and the downlink coverage (again refer to 'q' in Figure 6.4).

In this case a three-sectored site is assumed with a data service of 144 kbps, therefore using the previous link budget (refer to Table 6.4) for 144 kbps at a UE speed of 3 km/h, the uplink pathloss can be calculated. In the downlink an Eb/No of 5.5 dB and an orthogonality of 0.6 are assumed without any transmit diversity. The 'adjacent cell' to the 'own cell' interference ratio (i) is assumed to be a factor of 0.65. All cable losses have been taken into account and the assumed power output of the base station is again 10 W. Part of this base station power output has been allocated for the required downlink common control channels. The propagation environment will have a noticeable effect on the capacities in this example. Propagation environments refer to the type of terrain and clutter present in the specified coverage area.

As UMTS (3G) traffic will be asymmetric, there is different data usage for the uplink compared to the downlink and in the majority of cases, downlink traffic will be greater than uplink traffic. As we can see in the illustration Figure 6.5, a difference is achieved in the coverage if the downlink power is increased from 10 to 20 W. If an increase in the downlink power is achieved by 3 dB, then it will be possible to allow a 3 dB higher maximum pathloss, regardless of the load, to achieve the same coverage. As indicated in the illustration Figure 6.5, the coverage improvement is greater than the capacity improvement due to the load curve. If the downlink pathloss remains at 153 dB, which is the maximum uplink pathloss with a 3 dB interference margin, the downlink capacity can only be increased by 10 per cent (0.4 dB) from 680 to 750 kbps. Increasing the downlink transmission power is not considered to be an efficient method of increasing the downlink capacity. However, if the base station had 20 W of transmission power available, then splitting the downlink power between two frequencies would increase the downlink capacity (from 750 to 1360 kbps, i.e. 2×680 kbps $=$ 1360 kbps). This would give a capacity increase of 80 per cent but would have no significant effect on the coverage.

In summary, the splitting of the downlink power between two frequencies is a very efficient method of increasing the downlink capacity. However, for this scenario the operator would need to have sufficient frequency

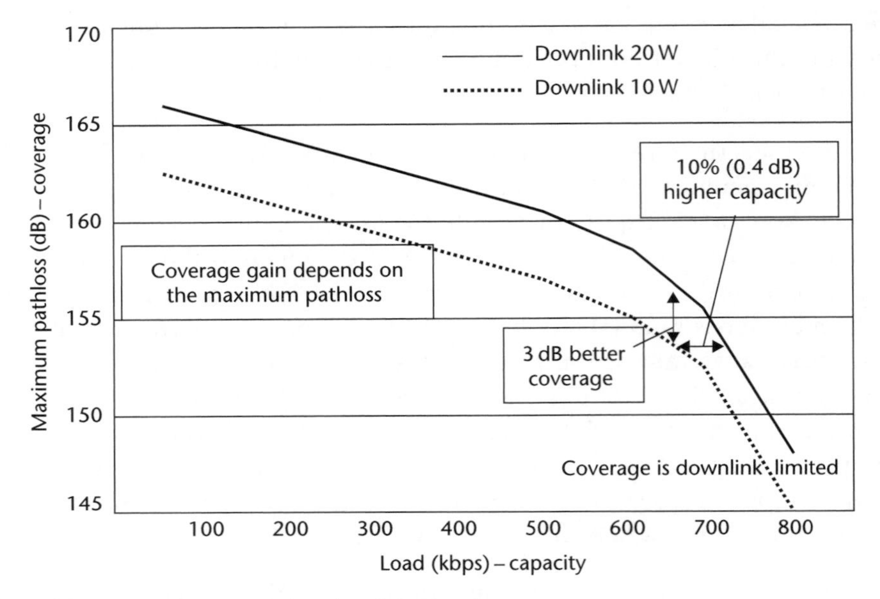

Figure 6.5 Coverage versus capacity – downlink comparison.

spectrum available, but the benefit would be a reduction in the requirement of the linear power amplifiers, a high investment aspect of the equipment in the base stations.

Finally, it should be noted that coverage is always downlink limited as it is more dependent on the load than in the uplink; this is because the maximum transmission power of the base station has to be shared between the downlink users. As previously mentioned, as each UE possesses its own power amplifier, even with a relatively low load occurring in the downlink, the coverage will decrease in proportion to the number of users within the cell, hence the drop-off in Figure 6.5 can be seen at a load of around 800 kbps.

6.4 Transmit Diversity

This subsection reviews methods of achieving further increases in RF gain by using transmit diversity. It will include both coherent combining gain and diversity gain. There are two different types of diversity present in the downlink: transmit diversity and multi-path diversity. Examples are given to show how fewer gains are achieved with downlink transmit diversity in contrast with uplink transmit diversity. The effect of transmitter antenna diversity should also be taken into account and included in the required Eb/No.

Finally this section reviews how transmit diversity gains can be used to increase coverage, giving examples on both capacity and coverage gains.

6.4.1 Downlink Transmit Diversity

The use of receiver diversity in the UE is unlikely to be implemented, mainly due to increased complexity in terminal design and consequent additional cost. It is considered far more viable to use a base station to implement transmit diversity, which is also known as space diversity. In order to achieve diversity the downlink signal is transmitted via two base station antenna branches. If receive diversity already exists in the base station, then it is possible to duplex the downlink transmission to the receive antennas, thus removing the need for extra antennas. Either polarization diversity or spaced antennas can be employed with downlink transmit diversity.

The performance gain from transmit and receive diversity consists of the 'coherent combining gain' and the 'diversity gain' against fading. Coherent combining gain is also known as beam-forming gain. This is the gain achieved from 'smart' antennas that are capable of beam forming. This is explained in more detail in Chapter 12, entitled Optimization through Detailed Site and Antenna Configurations. Diversity gain is literally the calculated gain that would have occurred had a diversity technique not been used.

When the RF signals are combined consistently together, coherent combining gain can be achieved, whereas interference is combined in a non-coherent manner. An ideal coherent combining gain of 3 dB could be achieved using two antennas. With regard to the uplink, the RAKE receiver is responsible for combining the signals coherently from the two diversity antennas. Dependent on the accuracy of the channel estimation, the coherent combining gain in the uplink is around 2.5–3.0 dB. It is also possible to obtain coherent combining in the UE, as shown in Figure 6.6, if both phases from the transmitting antennas are adjusted according to

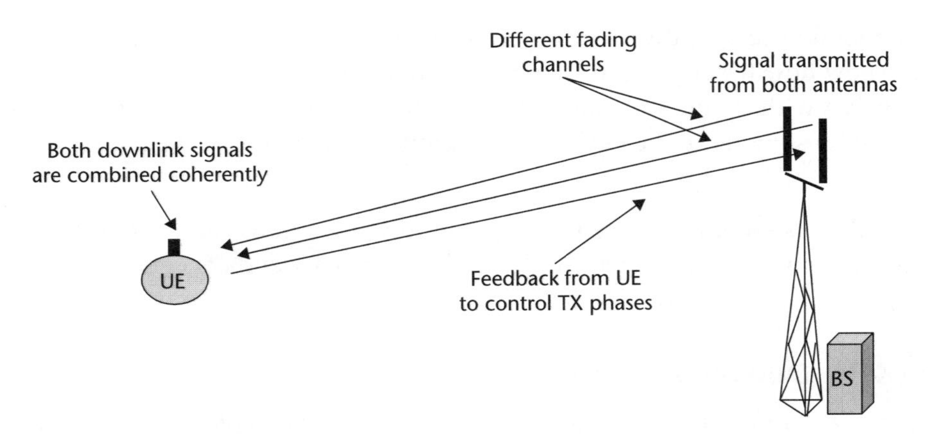

Figure 6.6 Downlink transmit diversity.

feedback commands sent by the UE. Only relative phasing of the signals received can be adjusted as the multi-path components cannot be coherently combined. The feedback loop is limited by a delay which is noticeable with high physical UE speeds (e.g. travelling in a car or train). This will also further degrade the coherent combining gain in the downlink. Therefore less gain can be achieved with downlink transmit diversity than with uplink receive diversity. However, both receive and transmit diversity provide an RF gain against fading that is therefore greater when there is less multi-path diversity present. Transmit diversity ensures that the downlink codes remain orthogonal in flat-fading channels (orthogonality is explained in Section 3.9.3) and alternatively, multi-path diversity causes a reduction in the orthogonality of the downlink codes. It is therefore beneficial to avoid multi-path propagation to ensure the orthogonality of the codes and therefore provide the diversity with transmit antenna diversity. This occurs, for example, when the UE is reasonably close with an unobstructed view of the base station which enables the interference-limited downlink capacity to be minimized, adding diversity in the form of transmit diversity, which is also considered advantageous. For further information regarding antennas please refer to Chapter 12.

6.4.2 Uplink and Downlink Transmit Diversity

As seen earlier in Table 4.8, comparisons were made between both the uplink and downlink capacities. Within these examples receive diversity was assumed for the uplink, however, no transmit diversity had been assumed for the downlink. It should be noted that in the downlink, the capacity gain can almost equal the capacity of the uplink in a macrocell environment. A comparison between the gains achieved in the uplink receive diversity and

<table>
<tr><th>Uplink receive (RX) diversity</th><th>Downlink transmit (TX) diversity</th></tr>
<tr><td colspan="2">Coherent combining gain
(ideal coherent combining gain = 3.0 dB with two antennas)</td></tr>
<tr><td>2.5–3.0 dB</td><td><2.5 dB</td></tr>
<tr><td colspan="2">Diversity gain against fading
Pedestrian 2.8 dB
Vehicular 0.8 dB</td></tr>
<tr><td colspan="2">Total gain in the reduction of TX power and the total gain from antenna diversity</td></tr>
<tr><td>3.0–6.0 dB</td><td>0.0–5.0 dB</td></tr>
</table>

Table 6.8 Uplink and downlink transmit diversity (based on WCDMA for UMTS, *Holma and Toskola, Wiley).*

the downlink transmit diversity can be seen in Table 6.8. This shows that the uplink receive diversity gains will be higher than the downlink transmit diversity, this is especially the case with the reduction of the TX power and the antenna diversity. Using these parameters listed in Table 6.8 for a sample microcell environment, capacities can be approximately equal without the use of transmit diversity. It should be noted that the interference-limited downlink capacity increases with transmit diversity and actually becomes higher than the uplink capacity within a microcell environment. As a general rule, the downlink requirements are likely to be higher than in the uplink and so the ability to achieve asymmetric capacity across the air interface is considered advantageous with this being attributed to the code orthogonality in the downlink. The transmit diversity gain could also be used to increase the downlink coverage, while the cell load remains constant. The coverage gain is higher than the capacity gain. However, with regard to coverage, if the uplink is the limiting factor, then it may not be possible to utilize the downlink coverage gains and extend the cell size by using downlink transmit diversity.

6.4.3 Capacity and Coverage Gains with Transmit Diversity

The results of the downlink transmit diversity RF gains which have been achieved with respect to the capacity and coverage, are shown in Figure 6.7.

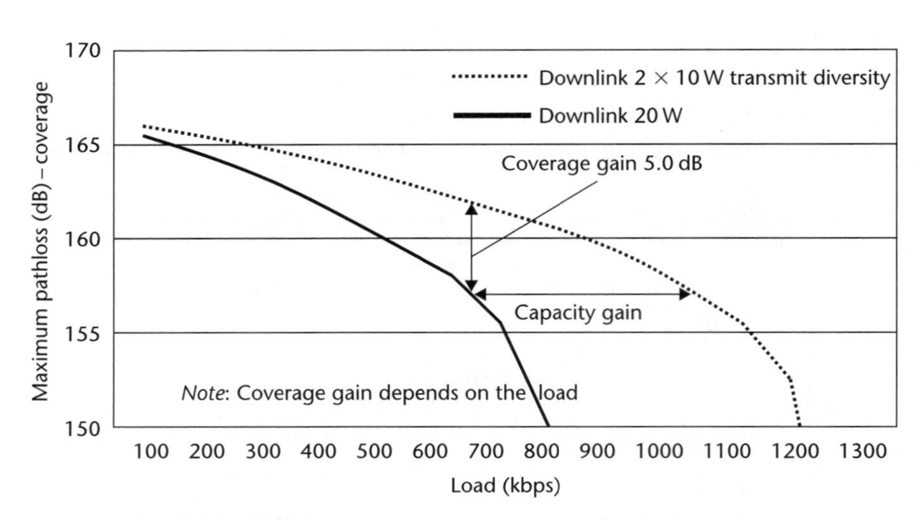

Figure 6.7 Capacity and coverage gains with transmit diversity.

This figure illustrates the difference between two antenna configurations, the default being a single antenna with a TX output of 20 W compared with two antennas utilizing transmit diversity and splitting the 20 W TX output between them. Both configurations have an assumed gain of 2 dB including both the diversity gain against fading and the coherent combining gain. For example, and again as indicated in Figure 6.7, if a maximum pathloss of 157 dB is assumed, then the capacity would be increased from 650 to 1030 kbps.

The downlink coverage can also be increased by using transmit diversity gain, while the load remains constant. In this example, the maximum pathloss could be increased by 5 dB from 157 to 162 dB if the load remains constant at 650 kbps. Therefore due to the load curve, it can be seen that the coverage gain is higher than the capacity gain. Unfortunately, as explained previously in this chapter (Section 6.4.2), if the uplink is the limiting factor regarding the coverage, it may not be possible to make use of any coverage gains that might be achieved by using downlink transmit *diversity* (refer to Section 6.4.2). However, any coverage gains that are achieved can be used to reduce the required base station transmission power. If, for example, the load remains the same as shown at 650 kbps and the maximum pathloss is unchanged and stays at 157 dB, it would then be possible to reduce the transmission power by 5 dB.

In summary it is necessary for the planner to be aware of both uplink and downlink transmit diversity, as this will prove beneficial from both a coverage and capacity point of view.

In conclusion, transmit diversity can be considered advantageous and can provide a gain against fading. It is important for the planner to be aware that generally speaking the downlink requirements are likely to be higher than the uplink requirements and therefore it is advantageous to attain asymmetric capacity over the air interface.

6.5 Range Calculation Parameters

This section considers the range calculation parameters (refer to Tables 6.9 and 6.10 for descriptions and examples of the parameters) taking into

Table 6.9 Range calculation parameters – descriptions.

	Parameter	Description
(a)	Average transmitter power per traffic channel	This can be described as the average or mean of the absolute maximum transmitted power occurring over a complete transmission cycle.
(b)	Cable, connector, and combiner losses at the transmitter	This is defined as the total losses of all the transmission system components between the antenna input and the transmitter output. All these losses will be positive dB values.
(c)	Transmitter EIRP/channel	This is defined as the summation of transmitter power output per traffic channel (dBm), transmission system losses (dB), and transmitter gain (dBi) in the direction of the maximum radiation.
(d)	Transmitter antenna gain	This is defined as the transmitting antenna's maximum gain in the horizontal plane. This will be specified in dB relative to an isotropic radiator (dBi).
(e)	Receiver antenna gain	This is defined as the maximum gain of the receiver antenna in the horizontal plane. This will be specified as dB relative to an isotropic radiator (dBi).
(f)	Cable, connector, and splitter losses at the receiver	This is the total of all the combined losses of all of the transmission system components between the receiver input and the receiving antenna output. All losses will be in positive dB values.
(g)	Thermal noise density	The thermal noise density plus the receiver noise figure is equal to the receiver noise density. In other words, this can be defined as the noise power per hertz at the receiver input.
(h)	Receiver noise figure	This is defined as the noise figure of the receiving system, reference to the system input.

(Contd.)

Table 6.9 Range calculation parameters – descriptions (Contd.)

	Parameter	Description
(i)	RX interference density	This can be defined as the in-band interference power divided by the system bandwidth. In-band interference power consists of both adjacent channel interference as well as co-channel interference. Hence, the RX and TX spectrum masks must be taken into account. RX interference density for the downlink is the interference power per hertz at the UE receiver, located at the edge of coverage. To summarize, the RX interference density is the interference power per hertz at the RX front-end.
(j)	Total effective noise plus interference density	This is defined as the logarithmic sum of the RX noise figure and the RX noise density, and the sum of the RX interference density.
(k)	Required Eb/(No + Io)	The ratio between the received energy per information bit, to the total noise and interference power density needed to meet the quality requirements. Based on the link level Eb/No value, where Eb is the energy per bit, and No represents thermal and receiver noise, the uplink range can be determined. The Eb/No value can be obtained from link level simulations.
(l)	Information rate	Channel bit rate in dB/Hz. Currently in UMTS the following non-packet options are supported: 4.75–12.2 kbps (AMR speech codec) and 32–384 kbps.
(m)	Handoff gain	This is the gain achieved by handoffs required to maintain the specific reliability at the periphery of the cell. The soft handover gain is acquired from the receiver, and is able to compensate for both slow and fast fading, as the signals arrive from almost uncorrelated paths. Typical values are assumed to be between 2.0 and 5.0 dB.
(n)	Receiver sensitivity	This can be defined as the signal level needed that fulfils the required Eb/(No + Io). An activity factor must be added to this value if real receiver sensitivity is required, for example if speech activity is required.
(o)	Other gain	Additional gains may be possible in the future due to new technologies. For example, space diversity multiple access (SDMA) may give additional antenna gains.
(p)	Log-normal fade margin	This is defined as the margin required to provide specific coverage probability, at the periphery of a single isolated cell. The propagation conditions are variable dependent on the RF environment, and hence the pathloss attenuation factor will also vary. The shadowing margin is dependent on the environment around the base station and the UE, and is directly related to the coverage requirement. A reasonable assumption for a log-normal fade margin would be around 10–13 dB. In this case it is assumed to be 11.3 dB.
(q)	Maximum pathloss	This is defined as the maximum pathloss that will still allow the required performance to be maintained at the cell periphery.
(r)	Maximum range	The maximum range is dependent on the actual range associated with the pathloss, and is calculated for each specific deployment scenario.

The values listed below depend on the type of propagation model used and should not be directly used for network dimensioning. With regard to different network solutions and services, the values listed here are all related, and can be used for both educational and comparison purposes.

Table 6.10 Range calculation parameters – sample.

Service	Medium bit rate data
User data rate	144 kbps
BER	10^{-3}
Delay	100 ms
Environment	Macrocellular
Cell layout	Hexagonal
Pathloss exponent	3.6
Log-normal shadowing	6 dB
Handover gain	5 dB
Fractional cell loading	70%
Base station antenna gain	6 dBi
UE antenna gain	0 dB
Maximum UE transmission power	1 W = 30 dB
Thermal noise	−174 dBm
Receiver noise figure	7 dB
Downlink capacity	169 kbps/cell = 7.0 users
Uplink Eb/No at constant (full) transmission power	4.0 dB

Source: WCDMA: Towards IP Mobility and Mobile Internet, Ojanpera and Prasad. (Sample range calculation table taken from code division test bed (CODIT)).

account the Intercell interference and propagation environment. The planner needs to be aware that the downlink may limit the range in a loaded network and hence more interference will be present (in the downlink). The factors required to achieve the goal of full coverage for lower data rates, the implementation of multi-user detection (MUD) and the requirements for achieving the optimal coverage for higher data rates are also covered.

The range calculation parameter values, listed below and described in the tables, cannot be given specific values as these will depend on the specific propagation model in use. However, these are covered in more detail in the Glossary section and some useful standard values are given. These parameters are used in the calculation for attaining link budgets:

a) Average TX power/traffic channel

b) Cable, connector, and combiner losses at the transmitter

c) TX antenna gain

d) TX effective isotropic radiated power (EIRP) gain

e) TX EIRP/traffic channel

f) RX antenna gain

g) Cable, connector, and combiner losses at the receiver

h) RX noise figure

i) Total effective noise plus interference density

j) Information rate

k) Required Eb/No (No + Io)

l) RX sensitivity

m) Handover gain

n) Other gain

o) Log-normal fade margin

p) Maximum pathloss

q) Maximum range

6.5.1 Cell Load and Range

The cell load has a marked effect on the achievable range, therefore if the number of subscribers increases, then the range decreases (see sample calculation in Table 6.10). As W-CDMA is an interference-based system, the system capacity is limited by the interference generated within its own arrangement. This is also evident within single cells in 3G, as opposed to orthogonal systems, such as time division multiple access (TDMA)-based networks.

6.5.1.1 Reduction in Cell Range

When the network load is increased, it is then possible to calculate the reduction in the cell range. However, the increase due to base station MUD must also be taken into account, with regard to the cell range, within a loaded network. Where a loaded network exists, having MUD implemented in the base stations, will enable the UE's transmission power to be reduced, thus making the range from an uplink perspective and also the average transmission power of the UE (approximately 125 mW), less sensitive to the actual network load. The planner must also take into account the base station intercell interference and the propagation environment.

When there is an unloaded network, the attainable coverage and range are limited in the uplink direction, as the total transmission power of the UE is always lower when compared to the transmission power of the base station in the downlink. The downlink may limit the range in a loaded network and therefore more interference will be present in the downlink than in the uplink.

6.5.2 Effects on Cell Coverage due to Parameter Modifications

There is a direct correlation between user data rate results and the reduction of the cell range. Therefore the planner's goal should be to ensure that full coverage will be available for lower data rates (below 144 kbps) while for higher data rates a limited coverage may be considered satisfactory. An Eb/No target of 4.0 dB has been assumed for the maximum ranges for different bit rates as shown in Figure 6.8. This takes into account all the range calculation parameters and services discussed previously in this chapter. Due to the benefits that can be achieved from implementing base station MUD (refer to Section 4.3), it should be considered as an upgrade solution to offer improved coverage. This is even the case when a high system load is present after the initial deployment due to MUD being able to reduce the UE transmission power in a loaded network. Thus a greater percentage of maximum load can be achieved with a network that has base station MUD implemented when the system is downlink limited. The performance of MUD must be taken into consideration to ensure good prediction and coverage planning. With MUD, good coverage does not need to suffer when a high system load is

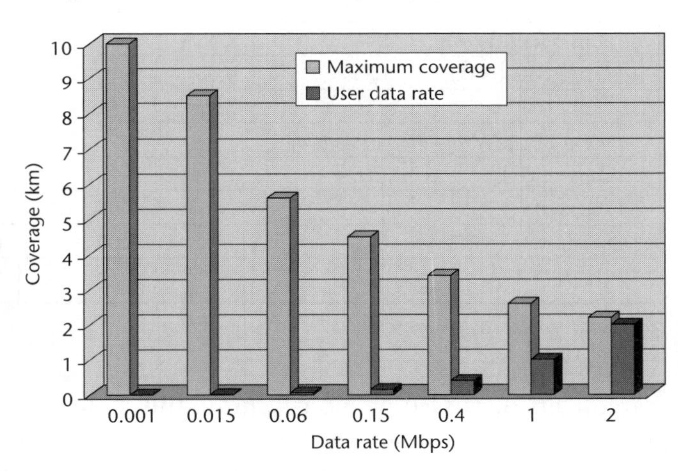

Figure 6.8 Coverage versus data rate.

present. It should be noted that an increase in the data rate will result in a reduced range in the uplink, however, the transmission power is unfortunately currently limited at the UE. This is why the coverage area for high data rate services, will differ from that of low data rate services.

6.5.3 Typical Coverage in 3G Universal Mobile Telephony System

Table 6.11 shows a sample of typical expected ranges in kilometres, based on calculations using the criteria shown in the range calculation

	Open terrain (km)	Rural (km)	Suburban (km)	Urban (km)	Dense urban (km)
Outdoor coverage with a 95% probability					
12 kbps voice	5.18	3.98	3.07	1.82	1.40
64 kbps data (RT)	4.48	3.45	2.66	1.57	1.21
144 kbps data (NRT)	5.25	4.03	3.10	1.84	1.42
384 kbps (NRT)	4.10	3.15	2.43	1.44	1.11
Indoor coverage with a 95% probability					
12 kbps voice	2.16	1.28	0.99	0.39	0.28
64 kbps data (RT)	1.87	1.11	0.85	0.34	0.25
144 kbps data (NRT)	2.19	1.30	1.00	0.40	0.29
384 kbps (NRT)	1.71	1.01	0.78	0.31	0.23

Table 6.11 Typical coverage in UMTS.

parameters discussed previously. From this table it can be seen firstly, that with a data speed of 384 kbps for outdoor coverage, with a 95 per cent probability, the range can vary from 4.1 down to 1.1 km which is a reduction of 73 per cent and secondly, that when comparing outdoor to indoor there is a range reduction of 58 per cent like for like. A further example would be to compare outdoor with indoor for a data rate of 144 kbps. With the outdoor coverage environment assuming a coverage probability of 95 per cent, a range of 1.84 km can be achieved within an urban environment. This can be compared to a similar data rate of 144 kbps within an indoor coverage environment assuming a coverage probability of 95 per cent, which would only achieve a cell range of 0.84 km within an urban environment.

6.6 Coverage Versus Different Bit Rates

In this section, associated issues of coverage and varying data rate usage are discussed. The Eb/No relationship, and the power link between the control channels and the data channels, which is paramount in maintaining a stable connection are discussed. Some power examples are given and finally issues with regard to overheads for varying data rates are discussed and explained with the aid of an example.

As discussed earlier in this chapter if higher data rate usage is present, then the processing gain is lower, hence the coverage will be reduced. In addition, with higher data rates the Eb/No is usually lower. With a lower Eb/No requirement the power needed to maintain the necessary performance can be reduced and hence the cell radius is increased. The required Eb/No is dependent on the data rate due to the fact that the dedicated physical common control channel (DPCCH) is responsible for maintaining the active connection as it contains the power control signalling bits and also the reference symbols required for channel estimation. Therefore Eb/No performance is dependent on the signal-to-interference ratio (SIR) estimation algorithms and the actual accuracy of the channel in question. The channel estimations are based on the reference symbols and thus will improve when more power is allocated to the DPCCH. Furthermore, the DPCCH does not contain any 'user data' as it is an overhead and should have its power minimized. It is therefore possible for the network to control the differences in power allocated to both the dedicated physical data channel (DPDCH) and the DPCCH. When there is data present in the DPDCH, the power is higher than that of the DPCCH. Some typical power differences between these two channels are listed with respect to different data rates in Table 6.12. This table has examples of four separate power levels for the two channels (DPDCH and DPCCH) with respect to four different data rates. For example, for a data rate of 384 kbps the power difference between the channels will be −9.0 dB.

Data rate (kbps)	Power differences (dB) expected between the DPDCH and DPCCH
1024 data	−12
384 data	−9
144 data	−6
12.2 voice	−3

Table 6.12 Coverage versus different bit rates.

There are 15 values available for differences in power between −23.5 and 0.0 dB and a 1 bit combination when there is no data transmitted, thus the DPDCH is empty, but the DPCCH must still maintain the current connection.

6.6.1 Overheads for Different Bit Rates

The illustration in Figure 6.9 shows the different data rates corresponding to the relative received power levels of the DPCCH, whereas the Eb/No

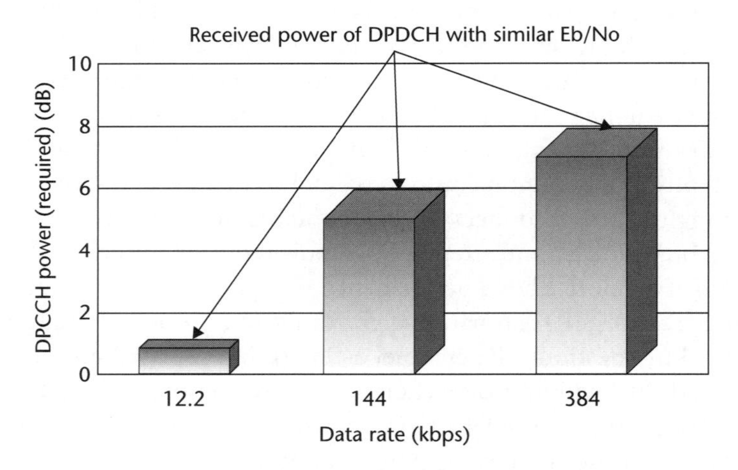

Figure 6.9 Control channel level power for different data rates.

value is assumed to be the same for all bit rates. As the received power is higher for the DPCCH with respect to higher data rates, the more accurate the channel estimation will be and the Eb/No performance is then also increased. As also illustrated in Figure 6.9, to maintain a data rate of 384 kbps through the DPDCH the power level of 8 dB will be required in order to maintain the control channel (DPCCH). However once the data rate of 384 kbps has been achieved the amount of overhead power required is only 0.5 dB, as shown in Figure 6.10. In contrast, a 12.2 kbps data rate requires only 1 dB to power the control channel but 1.7 dB in overhead power to maintain the channel. In summary the higher data rate requires more power for the control channel, but the overheads will be lower in comparison to a low data rate.

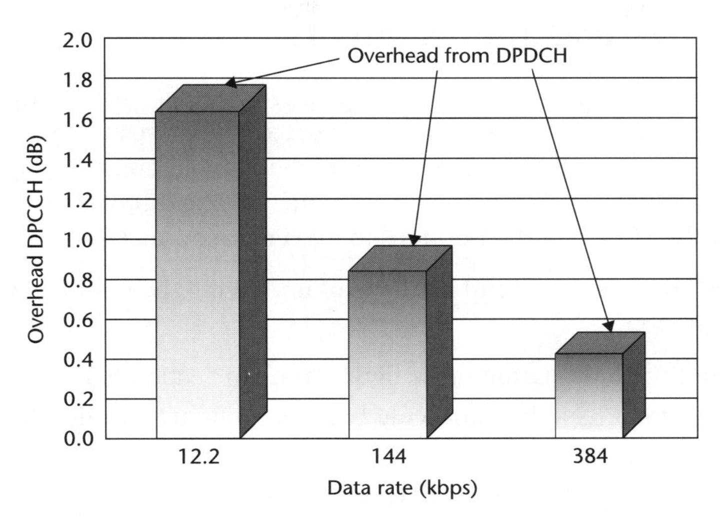

Figure 6.10 Overheads for different bit rates – Example 2.

Therefore when a high rate connection is active, then the power of the DPCCH is increased thus ensuring more accurate channel estimation and lower overhead for the DPCCH. Both the increased power and the lower overhead of the DPCCH contribute to improving the Eb/No performance. To achieve a wide area uplink coverage for the high data rate services will require accurate planning.

To achieve the goal of eventually providing two full Mbps uplink coverage will require a high base station density, particularly in urban areas. However, when the UE reaches its maximum transmission power, it is possible for the bit rate of the uplink transmission to be decreased during the connection and improve the coverage. This reduction would be possible for the AMR speech service and also for NRT packet data services which can tolerate delays.

In conclusion, the planner should be now aware of how different data rates will have an effect on the coverage area including the Eb/No relationship and the power link relationship between both the data and control channels. In addition, the overheads required for different bit rates have also been covered.

This leads into the final section which deals with coverage improvements.

6.7 Coverage Improvements

The final section in this chapter places greater emphasis on coverage and reviews certain issues that will enable coverage improvements to be achieved within the network. Reducing the base station noise figure, reducing the Eb/No, and reducing the interference margin, are also discussed with a view to giving the planner further options to maximize coverage.

Listed below are some useful methods of improving the uplink coverage of the base station:

a) Reduce the base station noise figure (refer to Section 6.5).
b) Reduce the Eb/No by improving base station baseband algorithms (refer to Section 6.2.4).
c) Increase the number of receiver antennas (refer to Section 6.5).
d) Increase the antenna gain (refer to Section 6.5).
e) Reduce the interference margin (e.g. the maximum capacity in the uplink) (refer to Section 6.2.1).
f) Reduce the cable loss between the low-noise amplifier (LNA) and the base station (refer to Section 6.5).

Reducing the base station noise figure can prove costly, as it will require improving all of the RF components of the base station receiver. The lower the noise figure, the better the system performance, as noise is an unwanted by-product that reduces system performance. To enable a reduction in the required Eb/No, the most important methods to be considered are to optimize the baseband algorithms in the base station, or alternatively to increase the number of receiver antennas, thus increasing diversity. The cell radius will increase in size if the required Eb/No value can be reduced, as there will be less power required to achieve the same performance. Increasing antenna gain can also improve the coverage and can be achieved by narrowing the horizontal RF pattern along with narrowing the vertical antenna beam. In addition, increasing the number of sectors present will also improve the signal gain. Adaptive antennas can also be used to achieve higher gains and are covered further in Chapter 12.

In summary for a three-sector site the maximum antenna gain likely to be achieved is around 18 dBi, with a vertical beam of 6°.

The main coverage improvements can be resolved by studying the coverage performance using a link budget. Some other areas to be considered are active antennas, mast-head amplifiers, increasing sectorization, and using repeaters. Other issues that should be taken into consideration consist of reducing the Eb/No along with the interference margin, reducing the base station noise figure, and reducing cable and connector losses.

6.8 Summary

This chapter has covered the main issues associated with coverage, breaking down link budgets along with the required specific parameters and providing some examples of expected coverage ranges. Interference and the all-important fading margin have been discussed, including soft handover gains, Eb/No, and the processing gain. Both coverage and capacity with relation to link budgets have been reviewed along with gains that can be achieved by the use of transmit diversity. Further topics such as range calculations along with some typical ranges for different bit rate coverage have also been given. Finally, utilizing the above information, other coverage improvements are listed.

Summary for Part III

This concludes the topic coverage and network planning. The planner should now be aware of the issues with respect to the second main parameter of the CCQ model, that of coverage.

Both 2G co-planning and co-existence along with all the associated areas involved in the planning and construction of link budgets including the range calculations and various improvements that can be made with regard to coverage have been reviewed.

Part IV consists of the third main parameter of the CCQ model: QoS and its importance to the planner.

PART FOUR

Quality and Network Planning

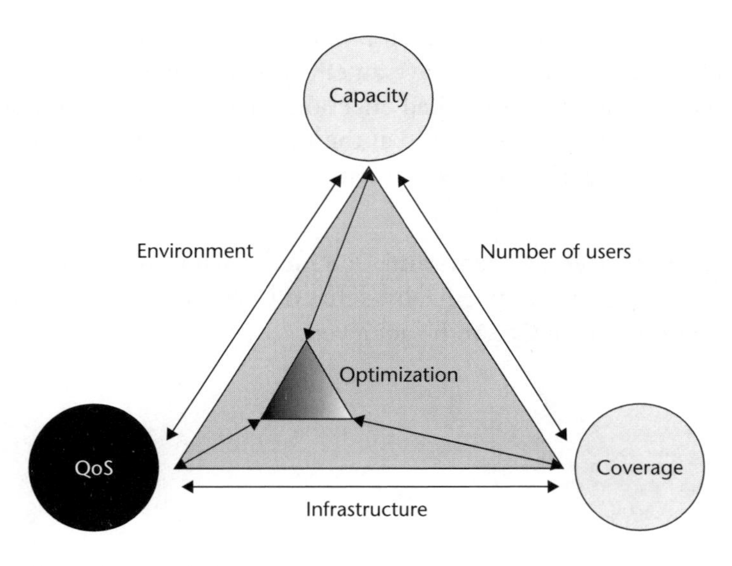

The CCQ model – QoS.

Introduction

The following section covers the third and final node on the coverage, capacity, and quality-of-service (QoS) model, abbreviated as CCQ model, that of quality and in this particular case the QoS. The third-generation (3G) services that will be utilized are considered to be 'end to end' or, alternatively, from one terminal equipment (TE) to another TE. For example, this will consist of the service being utilized from the end user's terminal, right across the entire network to the opposite end of the connection, in most cases the Internet. This 'end-to-end' service will possess a certain QoS, which is provided for the end user. In order that both low and high speed 3G services are able to operate correctly, the quality of the transmission must be guaranteed at the end user terminal. Therefore it is the particular service that is being accessed that will demand these levels of quality. To achieve a certain network QoS, it will be required for the specific QoS mechanism(s) being utilized, to establish and set up a bearer service with certain defined characteristics and functionality pertaining to the required service (e.g. video streaming, video to video, Internet browsing, and e-mail delivery).

As 3G network planning is based on predictions, and these can at times be inaccurate, this in turn will have an effect on the predicted data speed availability and therefore a marked effect on the QoS. If in the model capacity and coverage are favoured at the expense of quality then the completed 3G network could be left with the functionality of a 2G or 2.5G network.

In this section the requirements and QoS mechanisms available are reviewed, along with the specifically defined QoS traffic classes that will enable certain levels of QoS to be achieved.

Quality of Service

To introduce 'quality of service' (QoS), it can be thought of as the service performance that is required by the end user to enable the required service or application to function correctly, maintaining efficient service throughout the connection. This chapter briefly describes the QoS service capabilities, the different levels of QoS classes already defined and some of the possible mechanisms that can be used to achieve the required QoS. These standardizations as defined by the 3G partnership project (3GPP) are correct at the time of writing and are subject to change.

7.1 Quality of Service

The 3G services are likely to start with the continued use of voice services with data usage having a gradual penetration as the new networks mature. It is, however, predicted that data usage will eventually dominate the entire volume of traffic although when this is likely to occur is still difficult to envisage at this stage.

When the transition from voice to data does occur, the movement from circuit-switched services to packet-switched services will become evident. Initially, it is not expected that all the QoS functions will be implemented when the 3G universal mobile telephony system (UMTS (3G)) services begin. So delay-sensitive applications, such as real-time (RT) video, must be carried on circuit-switched bearers which have the ability to support delay-critical high data rate services. Due to the different delay-sensitive requirements of the various services that have evolved and will continue to evolve on the 3G networks, four different groups of traffic classes have been defined and these are covered later in this chapter:

a) Conversational class
b) Streaming class
c) Interactive class
d) Background class

These four classes take into account how delay-sensitive the required traffic is. A QoS level of '1' equates to the conversational class traffic and is very delay sensitive. Alternatively, a QoS level of '4' equates to the background class traffic and is very insensitive with regard to delays. As UMTS (3G) will have to support a wide range of applications, many of which are still being developed, they are all likely to possess different QoS requirements. An indication of some of the services that will be used within UMTS (3G) is illustrated in Figure 7.1 and shows the different levels of QoS required.

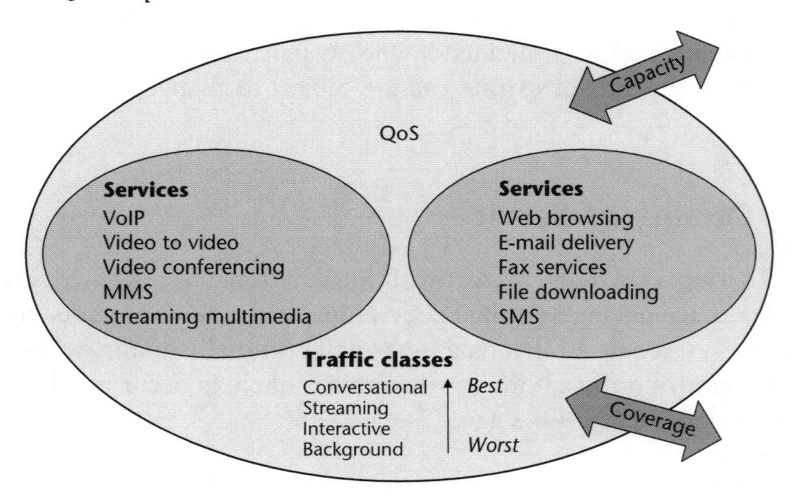

Figure 7.1 QoS examples.

At present it is not possible to predict the nature and usage of many of these applications and therefore, with this consideration, it is neither logical nor viable to optimize UMTS (3G) for only one type of application. However, once a connection is active, it will be possible for a user/application to negotiate the bearer characteristics required for carrying the specific data service thus allowing the planner some flexibility. In addition, during the course of the active connection, it will be possible via a bearer re-negotiation procedure to change the bearer, or in other words the data rates for the service in use. The actual application in use can initiate the bearer negotiation and either the network or the application can initiate a re-negotiation when and where necessary. For example, in the case of the network, this could occur due to handover situations and in the case of application-initiated bearer negotiations, the application will request either more or less data as required with the network checking that there are available resources to process this request, without disturbing the existing users. The user will then be able to either accept or reject the required changes, and the properties of the specific bearer will have a direct effect on the price of the service. This is covered in more detail in Chapter 9, entitled Radio Environments and Microcell Planning.

7.2 Service Capabilities

Unlike previous telecommunications technologies where the emphasis has been on finding the premium application, with UMTS (3G) a 'cocktail' of premium applications is likely. Various applications are already predicted to be widely popular, such as location-based services, gaming applications, and sport coverage. However, other services will also evolve and, with them, the abilities, expansion, and depth of the UMTS (3G) network that will develop and mature to fulfil the various service capabilities as shown in Table 7.1. Below are some examples of the possible 3G services that are likely to exist in the near future:

a) Location-based services
b) Virtual banking, currency downloading
c) Multimedia messaging
d) Video on demand, on-line library, and books
e) M-commerce
f) News and traffic flashes, video phoning
g) Gaming applications
h) Ticketing services, interactive shopping
i) On-line translations, intelligent searches

j) Desktop video conferencing, voice recognition, and response
k) Interactive and virtual school, universal subscriber identity manual (SIM) and credit card

These services will all run alongside the existing system currently available in the global system for mobile communications (GSM) networks.

	Peak bit rate	Maximum transfer delay	BER
RT services			
Outdoor (speeds up to 250 km/h)	144–384 kbps	20–300 ms	10^{-3}–10^{-7}
Urban/suburban/outdoor (speeds up to 150 km/h)	384–512 kbps	20–300 ms	10^{-3}–10^{-7}
Indoor/low range outdoor (speeds up to 10 km/h)	2 Mbps	20–300 ms	10^{-3}–10^{-7}
Non-RT services			
Outdoor (speeds up to 250 km/h)	144–384 kbps	150 ms or more	10^{-5}–10^{-8}
Urban/suburban/outdoor (speeds up to 150 km/h)	384–512 kbps	150 ms or more	10^{-5}–10^{-8}
Indoor/low range outdoor (speeds up to 10 km/h)	2 Mbps	150 ms or more	10^{-5}–10^{-8}

Table 7.1 QoS capabilities.

7.3 Quality-of-Service Traffic Classes

For all UMTS (3G) bearer services four traffic classes have been defined. The determining factor of these classes is the difference in the delay sensitivity; this will therefore determine which type of service (TOS) or application can be used with the respective QoS class. The first class definition known as the 'conversational class' has very sensitive delay requirements and will be used for RT traffic, as opposed to the opposite end of the scale, the 'background class', which is the least sensitive to any delay.

7.3.1 Conversational Class (Optimum Real-Time Quality of Service)

For RT speech, the maximum end-to-end delay must be less than 300–400 ms. Voice traffic is generally speaking symmetric and within

UMTS (3G) an adaptive multi-rate (AMR) speech codec will be employed. This enables eight source rates to be available from 12.2 down to 4.75 kbps and is beneficial because it can be used to reduce the amount of data required for speech, therefore freeing up resources for other users.

7.3.2 Streaming Class

The streaming class can guarantee a bit rate over the UMTS (3G) network, thus allowing the use of services that are jitter sensitive. A relatively high QoS level is still required for this, as a continuous and steady stream of data must be processed. The user is then able to start displaying the data service before the entire file is transmitted. Delays can be tolerated here as streaming applications are very asymmetric and are able to withstand greater delays than symmetric speech services; furthermore any jitter occurring within the transmission can also be tolerated with streaming. This can, however, be smoothed out by a jitter buffer, whose function is to ensure the packets are scheduled in a manageable fashion into the packet disassembler.

7.3.3 Interactive Class

A good example of interactive class traffic is web browsing, as this type of traffic can be characterized by the end user's request. As a request occurs, the end user is expecting a response within a certain time, but delays here are acceptable as there is a round trip delay time to be taken into account. Finally, the content of the packets must be transferred transparently with a low bit error rate (BER). Other examples could be via remote-based equipment and would consist of database retrieval and server access.

7.3.4 Background Class

With background class, delays of seconds, tens of seconds, or even minutes may occur. As delays in this class are not critical to the required service the data does not need to be transferred transparently and will then be received error free. Examples of applications suitable for use in the background class are e-mail delivery, short message services (SMSs), database, and file downloading.

7.4 Quality-of-Service Mechanisms

QoS mechanisms are required to ensure the requisite QoS can be both achieved and maintained for the duration of the end-to-end connection. As this end-to-end connection may consist of a connection established between the user equipment (UE) and the Internet, the data connection, for example, may pass through many different routers before reaching its final destination. Therefore, QoS mechanisms are required in order to ensure the necessary bandwidth for this end-to-end connection is provided, otherwise an unacceptable reduced level of QoS will occur.

There are various mechanisms available, although there does not yet exist a perfect single mechanism, for performing this complex function of bandwidth reservation upon demand. The main mechanisms, covered below, discuss the current protocols available (at the time of publication) including H.323, session initiation protocol (SIP), real-time protocol (RTP), and reservation protocol (RSVP) which control the QoS.

Initially the QoS levels of conversational and streaming classes will be transmitted via circuit-switched connections, in RT over the wideband-code division multiple access (W-CDMA) air interface. Alternatively, the lower QoS levels comprising the interactive and background classes will be transmitted as scheduled non-RT packet data. The inherent problem exists in defining QoS levels within the Internet protocol (IP). The QoS at present within IP is classed as 'best effort' and this causes problems with high data rate and delay-sensitive services as the QoS cannot be guaranteed using the existing packet technology. There are mechanisms available, but constructing a mechanism that can guarantee an end-to-end connection is somewhat challenging. Standard video telephony is based on the H.323 protocol (as defined by the International Telecommunications Union (ITU)) which, unfortunately, is not the best solution to be used through the Internet and particularly across an air interface, due to the complex control signalling required. RT communications across the Internet can be provided by the Internet Engineering Task Force (IETF) RTP set, which allows the initiation of a fixed bandwidth for the end-to-end link. Using the RSVP allows sufficient bandwidth to be reserved, however, this requires that every router along the way must support RSVP for end-to-end bandwidth reservation to succeed. It should be noted that in practice this is complicated and is not considered realistic for use within the Internet router network, because the reservation of bandwidth would cause severe disruption throughout the Internet. In addition, both scalability (in that it can be increased or reduced without loss of form) and billing will be problematic with RSVP.

The RTP can be considered as another protocol for achieving QoS and is determined by the IETF. With RTP the data delivery is monitored by means of a closely integrated control protocol called real-time control protocol (RTCP). However, RTP does not provide any mechanisms to ensure timely delivery or provide QoS guarantees.

Another method is to use the SIP, which is addressing neutral. This means that it ignores IP addresses, with addresses expressed in uniform resource locators (URLs). SIP is also independent of the packet layer and requires only an unreliable datagram service as it provides its own reliability mechanism while the others are not designed to have a reliability mechanism. SIP is able to provide the necessary protocol mechanisms for services such as call forwarding, call forwarding-no answer, call forwarding-busy, other address translation services, caller and recipient number delivery, independent address even when the user changes terminals, and finally, terminal-type negotiation and selection. The last main advantage with SIP is that the callers can also be presented with a choice of how to establish the required connection, for example, via Internet telephony, standard mobile phone or answering service, to name a few.

Another approach would be to use the Differentiated services mechanism (known as 'Diffserv'), which is more 'IP like' but cannot guarantee the quality to the level of RSVP. However, this protocol is based on a 'connectionless' approach and 'marks' each packet's TOS, or priority bit, which can be interpreted by the routers. Nevertheless, Diffserv still requires a mechanism for scheduling and queuing in accordance with the required QoS product. Diffserv mechanisms can provide a predictable performance for a certain load at a specified time and can provide a simple priority control by re-defining the TOS byte within the IP header. However, this will not provide an on-demand QoS request and therefore Diffserv cannot guarantee the required QoS as it is ultimately based on a connectionless approach.

In summary, there is no preferred choice of protocol which will give the optimal QoS, as there are pros and cons for each specific protocol. It is likely that, as the infrastructure matures, the long-standing issues involved with QoS will be resolved as newer methods and protocols become available.

The planner should now be aware of the issues involved with QoS and how to dimension a network accordingly. The four all-important traffic classes of QoS have been covered and the current QoS mechanisms have been discussed giving the reader a valuable insight into the complexity of QoS and how it can be achieved. However, we have also learnt that achieving the required QoS will prove challenging, especially with regard to RT high data rate usage, and have surveyed the difficulties involved in achieving the ultimate balance of CCQ.

Radio Resource Management

8.1 Introduction

Radio resource management (RRM) handles the utilization of the air-interface resources, thus ensuring that the data throughput of the radio access network (RAN) can be efficiently managed, for example management of the power, spectrum, and channels available. This will in turn ensure that the variety of data rates and different services can be maintained with the required quality of service (QoS). With regard to the coverage, capacity, and QoS (CCQ) model, the RRM is responsible for providing the optimal trade-off scenario between the three key parameters of CCQ.

RRM can be divided into different functionalities pertaining to the corresponding network elements as shown in Figure 8.1. Power control is

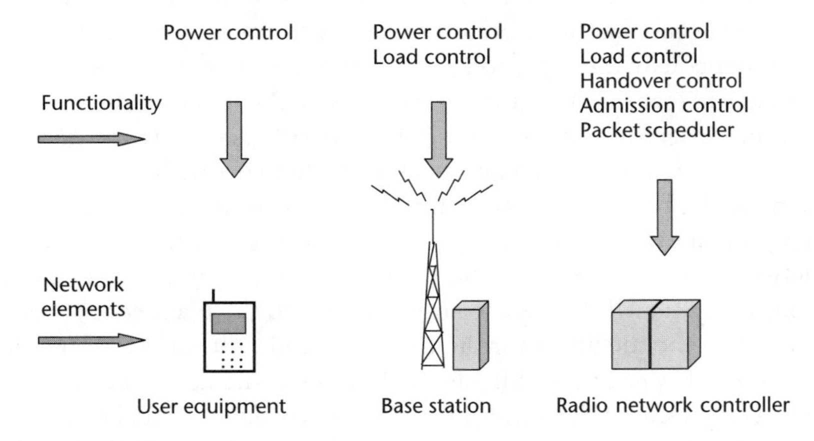

Figure 8.1 Radio resource management.

one of the important factors to be taken into consideration in a wideband-code division multiple Access (W-CDMA)-based system as its main function is to reduce the interference levels in the air interface as much as possible, and, in doing so, provide the required QoS.

RRM is responsible for managing the entire system's data throughput in the most efficient and timely manner. As there will be a variety of different services along with different QoS requirements and different data rates, RRM algorithms will be required for the admission control, load control, and packet scheduling.

Finally, with regard to RRM, two types of blocking must be taken into consideration: hard and soft blocking. Hard blocking occurs when the hardware is the reason for limiting the systems' capacity, before the air interface becomes overloaded. Soft blocking is defined, whereby the air-interface load is estimated to be higher than the pre-planned limit. In general, soft blocking-based RRM will allow a higher capacity than with hard blocking RRM. However, the air-interface load should be measured first, if soft blocking is to be applied. All issues relating to RRM are covered in further detail throughout this chapter.

8.2 Admission Control

There are a number of limiting factors with W-CDMA radio access, some of them dependent on the radio environment and others of a fixed nature. However, what must be considered as the most important factor, and simultaneously the most complex to manage and control, is the interference present in the air interface. As W-CDMA is run on interference, every user equipment (UE) currently accessing the network is seen as a source of interference by all the other UEs within that sector or cell. One of the basic and most important criteria to be considered, when planning the network, is defining an acceptable interference level, whereby the network will be able to function within the required limits. The signals transmitted by the UE across the air interface and the planned interference values will define the limits for the capacity within the air interface. The signal-to-interference ratio (SIR) is used in this context, and can be defined as the ratio of the amplitude of a desired signal to that of a signal interfering with its reception. So the network must theoretically be able to withstand a maximum SIR at a certain level in one cell, therefore there will be a difference in the power levels between both the signal and the interference to enable one signal (code) to be extracted from all the other signals, simultaneously transmitting on the same carrier. If there is only a negligible power difference between the interfering components and the required signal (code), the base station will be unable to extract this required signal from the carrier. A part of the SIR is 'consumed' by every UE maintaining an active radio connection, therefore it can be stated that the cell will be operating at its maximum level, at the point when the base station is unable to extract the required signal(s) from the carrier.

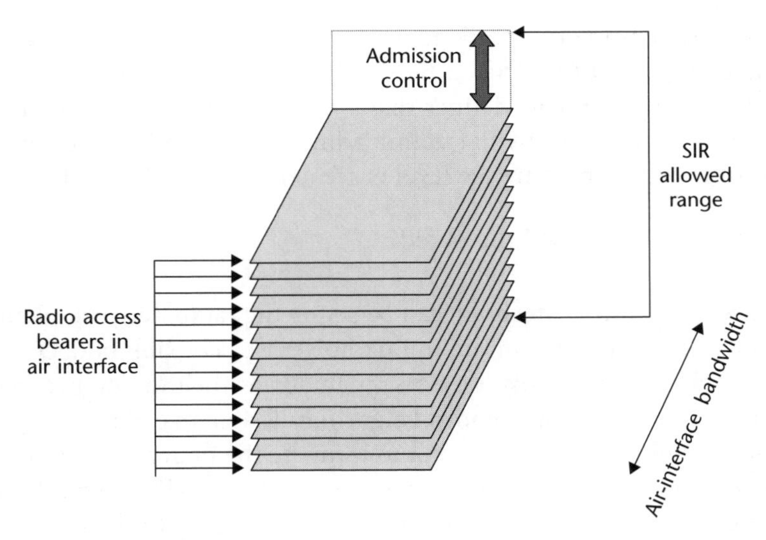

Figure 8.2 Admission control.

A graphical explanation of admission control is illustrated in Figure 8.2. As can be seen from the illustration, the number of radio access bearers that can be allowed to access the RAN is limited by the SIR. This range will continuously be varying and will be dependent on the cell load and the data rates being used by the current users within the cell. The main function of admission control is to estimate whether a new radio bearer can be allowed to access the system without sacrificing the radio bearer requirements of the existing users. Therefore, the admission control algorithm must be able to predict the cell load, assuming that the new radio bearer is admitted into the system. In addition, it should be noted that the terrestrial transmission resource availability is also verified. The radio network controller (RNC) will either allow or reject access to the network based on the admission control. This leads to the following subsections.

8.2.1 Admission Control Strategy

The admission control strategy describes the parameters involved with regard to the cell threshold, followed by the congestion- and interference-driven levels of admission control.

The basis of the primary uplink admission control decision criteria can be defined as the use of total power received at the base station. The admission control strategy defines that a new radio bearer will not be admitted to the network by the uplink admission control algorithm, if the new resulting total interference level is greater than the threshold value:

$$I \text{ total\_old} + \Delta I > I \text{ threshold} \tag{8.1}$$

The threshold value 'I threshold' can be set by the radio network planner, and is identical to the maximum uplink noise rise. The link budget for the specified cell must have this noise rise included as the interference margin, hence the uplink admission control algorithm is able to estimate the load increase. The objective is to attempt to estimate the increase 'ΔI' in the uplink received wideband interference power 'I total', which would occur assuming that a new radio bearer (or user) wishes to enter the network. The power increase estimates and the admission of the new radio bearer are both handled by the admission control functionality. The admission control allows the following three definable parameters to be limited by the user:

a) Uplink noise rise in the network (I threshold).
b) Downlink load or the downlink TX power.
c) Maximum TX power per user.

When considering macrocells, a typical maximum noise rise of 6 dB (which would equate to a load of 75 per cent) is permitted. For microcells, a 20 dB noise rise is considered acceptable. Noise rise is defined as the increase in wideband interference over the thermal noise in the base station reception. Less noise rise is allowed in macrocells, as the reduction in the coverage area is more significant than for microcells. The illustration in Figure 8.3 indicates the uplink load curve, and the estimated increase in load due to a new radio bearer entering the system.

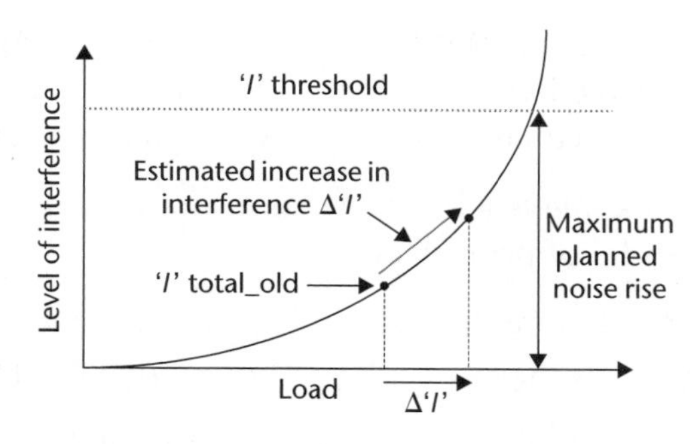

Figure 8.3 Uplink load curve and estimation of load increase due to a new user.

8.2.2 Levels of Admission Control

There are two levels of admission control which can be applied within a universal mobile telephony system (UMTS) network. These consist of congestion-driven and interference-driven admission controls. The congestion-driven admission control interacts with the transmission control protocol/Internet protocol (TCP/IP) stack, resource reservation protocol (RSVP)/multi-protocol label switching (MPLS), and differential service (Diffserv) traffic shaping protocols. The interference-driven admission control interacts with the media access control (MAC) physical layer protocols, and is effectively an asynchronous transfer mode (ATM) available bit rate (ABR) admission protocol. More in-depth information on these protocols can be found in *WCDMA: Towards IP Mobility and Mobile Internet* by Ojanpera and Prasad (Artech House, 2001). When considering the properties of heavy-tailed traffic loads, the impact of these levels of admission control is still incompletely understood. It is, however, known that control optimization will be critical in delivering a consistent and varied media, as will be required by the end users. This is now beginning to become fully acknowledged, and is currently undergoing further research and development.

8.3 Load Control (Congestion Control)

8.3.1 Load Control Actions

Load control must be considered as a vital function of RRM to ensure that the system remains stable and is not overloaded. If the network has been correctly planned, system overload occurs rarely. However, should an overload scenario occur, then the load control functionality is responsible for rectifying the overload, and returning the network to the originally planned target load in a timely and efficient manner.

Possible load control actions for reducing the system load are listed in order of priority below:

a) Ignore the downlink 'power-up' commands received from the UE.
b) Reduce the uplink Eb/No target used by the uplink fast power control.
c) Reduce the packet traffic throughput.
d) Handover to an alternative W-CDMA carrier.
e) Handover to a second-generation (2G) system.
f) Decrease the data rates of real-time users (e.g. adaptive multi-rate (AMR) speech codec).
g) Drop calls in a controlled manner.

These have the effect of both reducing the Eb/No target used by the uplink fast power control and ignoring the downlink 'power-up' commands from the UE, which can be executed from within the base station and can assist in providing for fast prioritization of the varied requested services in use.

When striving to maintain the QoS required at a high data rate, it should be noted that real-time services that employ a high data rate along with a high QoS level may be impaired if they have to be re-transmitted, for example, due to all the signalling required. Lower data rate packet services will suffer with further increased delays, while the quality of the delay-sensitive services, such as speech and video, are maintained. All other load control actions previously mentioned will be slower.

The packet scheduler is used to reduce packet traffic. It is possible to reduce the data rates of any users currently maintaining real-time speech connections via an AMR speech codec, as lower source rates can be utilized. Load control algorithms, inter-system handovers, and inter-frequency handovers can be used to balance traffic loads. In the worst case scenario, it may be required to terminate real-time users, thus rapidly reducing the load so as to return the system to a manageable level. However, as there are a number of afore-mentioned possibilities to reduce the system load, the dropping of real-time connections is always considered as a last resort, and should not occur within a well-planned system.

8.4 Packet Scheduler and Load Control

The packet scheduler is considered as an important part of the load control functionality, hence they are both closely connected. The load of non-real-time connections is somewhat easier to manage and control, as the packet scheduler is not required to guarantee any delay experienced with the non-real-time connections. It is always advantageous for the packet scheduler to decrease the controllable load of the non-real-time users to compensate for the uncontrollable real-time connections as illustrated in Figure 8.4. With the use of advanced packet scheduling algorithms, the system can be maintained at the required level. It is necessary for the admission control to perform an estimate of the continually changing load created by the non-controllable real-time users. This load must not be disrupted and will remain in an undisturbed state at the expense of the controllable load. Therefore, the controllable load refers to the non-real-time users, as their requirements can be decreased as and when required. A typical example occurs when a video-to-video connection is requested; then the admission control must estimate how much controllable load will have to be reduced or sacrificed in order to admit the real-time video user to the system, while simultaneously ensuring that the load is maintained within the defined target load parameters. The available data rates that can be assigned to a new connection and the connection setup parameters are also defined by the admission control, as described previously in this chapter. The ultimate aim of the packet scheduler and load control is to manage both the controllable and non-controllable load within the cell.

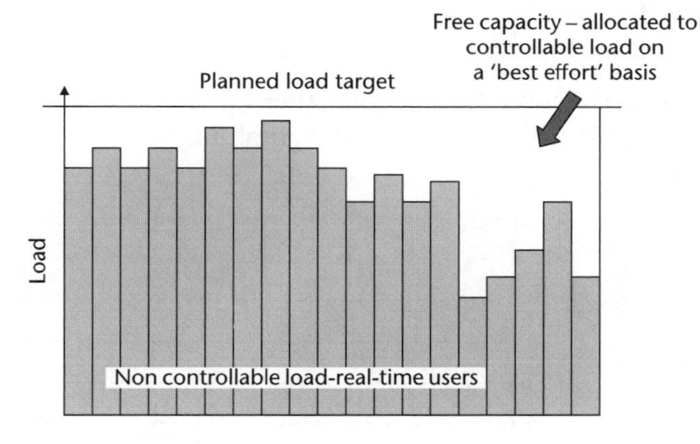

Figure 8.4 Packet scheduler and load control.

8.5 Interference Margin

The cell load has a direct relationship with the SIR or the interference margin. The SIR can be thought of as the ratio of power present in the required signal, to the interference power residing in the channel. If the cell load is expressed as 'load factor' (e.g. from 0 to 1 equals cell percentile load) and the interference margin is expressed as 'I', this leads to the equation shown below:

$$I = 10\log\left(\frac{1}{1 - \text{load factor}}\right) \tag{8.2}$$

Taking different 'load factor' values, the interference margin calculations can be placed together and the graph illustrated in Figure 8.5 can be

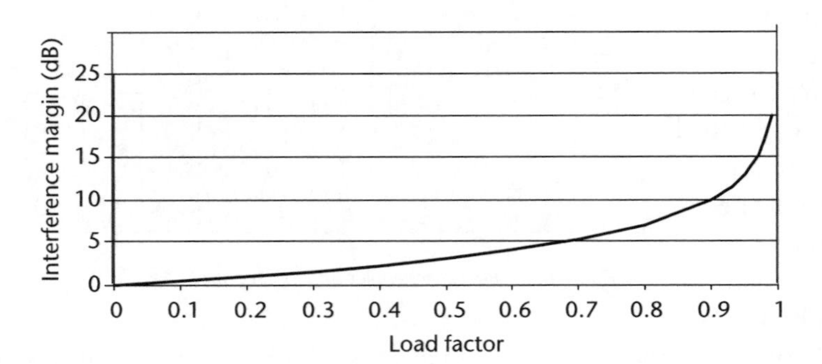

Figure 8.5 Interference margin and load factor.

plotted. As can be seen from the graph, the interference will be very difficult to control once the cell load begins to exceed 70 per cent, as the curve starts to go exponential. This is an important factor to be aware of, as W-CDMA networks are interference based, thus it is recommended to dimension the system to only allow a 'load factor' value of 0.5 (50 per cent) of the predicted traffic capacity. This value ensures that a safety margin exists and will guarantee that the network behaves in a stable manner. It is expected in certain cases that this 50 per cent value will be exceeded as network operators, hungry for extra capacity, attempt to maximize the traffic load to its limit. However, this is not recommended, as a situation could transpire whereby the load becomes too high and the system becomes unstable. At this point, the RNC is no longer able to maintain control and cell breakdown can occur resulting in a loss of all traffic within the cell in question, while simultaneously increasing the load of all adjacent cells, thus causing further disruption.

8.6 Power Control

Power control is an issue that needs to be understood, as it is essential within the W-CDMA RAN. Accuracy is critical and will have a marked effect on the overall system capacity and performance, unlike previous TDMA-based 2G networks. The following three bullet points list the reasons why the capacity and performance are affected within a W-CDMA network:

a) Transmission from the UE occurs simultaneously in time, and not in different timeslots as with 2G global system for mobile communication (GSM) systems.
b) The W-CDMA RAN has a frequency re-use factor of one (it is possible to utilize other frequencies as well, assuming that additional blocks of the frequency spectrum have been purchased).
c) Any inaccuracy in the power control mechanisms will immediately result in increased interference, therefore decreasing the capacity of the network.

The power transmission levels of all UEs within one cell must be seen as equal regardless of their location within the cell as illustrated in Figure 8.6

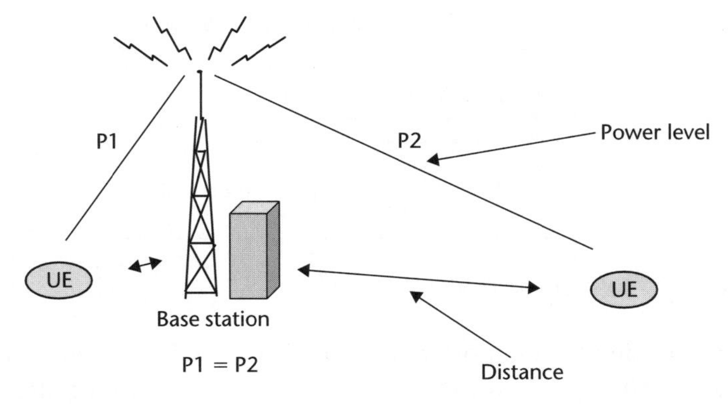

Figure 8.6 Power control.

in which the power level of 'P1' must be equal to 'P2'. With respect to the distances and radio paths involved, a similarity does exist with GSM, however, due to the three reasons already stated, that the power control mechanisms must be able to act both fast and accurately to ensure system capacity and stability. The optimal situation will occur when the power levels of all UEs are seen as equal by the base station receiver, regardless of the distance between the UE and the base station. If this situation can be maintained, this will ensure that an optimal SIR exists, thus allowing the base station receiver to decode the maximum number of UE transmissions. As this is considered a challenging task, the power control mechanisms employed in 2G GSM systems will not suffice, and are clearly inadequate for a W-CDMA system. Therefore, the following sections cover the power control mechanisms that are employed within a UMTS W-CDMA system.

8.6.1 Power Control Mechanisms

There are three power control mechanisms employed by a W-CDMA network and they are illustrated in Figure 8.7. Fast and accurate power

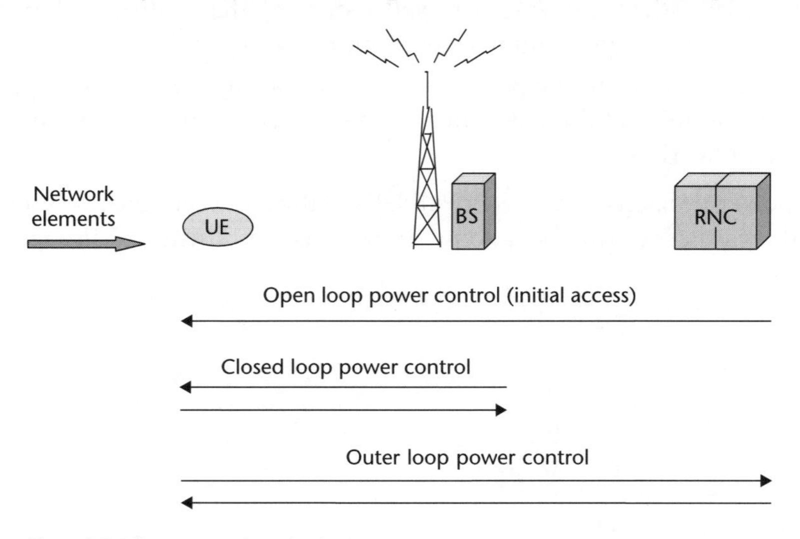

Figure 8.7 Power control mechanisms.

control is considered an important aspect within the RAN, and even more so with regard to the uplink. It can be stated that if power control did not exist, it would only take one solitary, over-powered UE to block an entire cell.

For example, if a scenario is assumed, whereby UE 'x' and UE 'y' are both operating on the same frequency, the data contained in their signals is only recoverable by their respective spreading codes. In this scenario, UE 'x' is located at the periphery of the cell, and incurs a 75 dB pathloss, while UE 'y' is located close to the base station. With no method of controlling the UE differing power levels to ensure the base station 'sees' the same power from both UEs, UE 'y' could easily 'swamp' UE 'x' and thus cause a large area of the cell to be blocked. In CDMA, this is known as the 'near–far problem'. Therefore to ensure that maximum capacity can be achieved, the goal from a power control perspective is to ensure that the received power per bit of all UEs is continuously equalized. Hence, taking the scenario just described, it is important for the planner to be aware of the following power control mechanisms described in the next subsections of this chapter.

8.6.1.1 Open Loop

The first power control mechanism to be discussed is classed as open loop power control. To enable a UE to access the RAN, an attempt must be made based on an estimate of the power level required. The UE is continually 'listening' to the network while it is in idle mode. The UE is able to periodically 'wake up' due to paging indicators received on the paging indicator channel (PICH) and then by measuring the received signal strength using automatic-gain-control (AGC) circuitry. This enables an estimate to be made of the propagation loss required to reach the base station. Additionally, the UE monitors the broadcast channel (BCH) and is then able to evaluate the pathloss with respect to the figures contained in the BCH channel. Based on these few measurements, the UE is able to estimate a required transmit power level needed to request access for a connection to the base station. This access request is sent as a burst on the random access channel (RACH), and assuming the estimate results were accurate enough, the base station will respond by sending a response burst on the acquisition indicator channel (AICH), confirming the reception of the RACH signature sequence. It is upon this reception of the AICH that the UE is aware that it has transmitted the previous burst at the correct power level, which will be identical to all other users currently maintaining an active connection within the cell. If the base station does not receive the original RACH burst, the UE will run through a 'time-out' sequence and initiate the whole process again. However, on this occasion the RACH burst will be at an increased power level of 1 dB, with respect to the previous burst. This procedure will continue, and the power level of the RACH burst will be increased in steps of 1 dB until the base station is able to 'hear' the UE in question. The universal terrestrial radio access (UTRA) specifications define the relative accuracy for a 1 dB power control step to be ±0.5 dB, however, it is likely that 'true' step sizes less than 1 dB will be somewhat complex to implement.

8.6.1.2 Closed Loop

Once the connection is established, the entire contents of the RACH are received by the base station, and closed loop power control is then initiated. This power control mechanism is tightly controlled, and has the ability to command the UE to either increase or decrease its transmission power level at the rate of 1.5 kHz (1500 times per second), which is operated on a one command per slot basis. Based on the SIR received by the base station, a decision is then made whether to increase or decrease the power. The SIR target is set by the outer loop power control described in the next section. The basic step size is 1 dB, however multiple power steps can also be performed as well as emulate smaller power steps. In the case of emulation, a 1 dB step for every second slot is performed, therefore emulating a 0.5 dB power step. Closed loop is also able to compensate for any fast fading, whereby no correlation exists between the uplink and the downlink. The basis of this process consists of firstly measuring the SIR in the base station. Upon evaluation of these measurements, the base station commands the UE to increase or decrease its transmission power. Therefore, after the SIR has been evaluated, it is compared to a set value known as the 'SIR set'. This means that if the current SIR is greater than the previous 'SIR set', the transmitter power control (TCP) commands the UE to decrease the power to 'down', otherwise it will increase the power to 'up'.

8.6.1.3 Outer Loop and Inner Loop

The bit error rate (BER) and the frame error rate (FER) of the service in operation must be taken into account since the outer loop power control is responsible for both setting and adjusting the Eb/No target with respect to the achieved BER/FER. It may also be possible to set the sample period and the measurement frequency. After the outer loop is set, the inner loop power control will be initiated and is responsible for achieving and maintaining the target Eb/No over the air interface. There are two algorithms available for use by the inner loop; these have the ability to control either a set or a single power control bit, which are dependent on the radio environment. The RNC must continually monitor, and be aware of, the current radio connections and their quality. This is especially the case where macro diversity is concerned, as the UE is in a state whereby it is simultaneously attached to the network via more than one cell. The RNC uses this inner loop mechanism to adjust the SIR of the connection. The effect can be powerful, as the network is able to compensate for any slow fading or any other changes occurring within the radio environment, with the end result that the target quality for the radio connection can be achieved.

8.6.1.4 Outer Loop

The outer loop control mechanism can be considered as the fourth mechanism and is located in the RNC. This mechanism initiates and defines the SIR parameter value required to be set, with respect to the FER. When an increase occurs in the FER, the value of the pre-defined '(SIR) set' is configured 'up' by the outer loop control. However, if there is no FER increase, then the '(SIR) set' is configured 'down'. The base station receives the 'SIR set adjustment commands' from the RNC in the downlink at rates varying from 100 ms up to several seconds.

Taking the uplink into consideration, the same procedure is initiated for the UE, which can be considered as the 'normal' uplink power control. Again, the new '(SIR) set' will be compared with the measured SIR value received by the base station. Assuming that the SIR is greater than the already '(SIR) set' value, the UE will be ordered to set the power to 'down' by the base station, otherwise the UE will set the power to 'up'.

An indication of outer loop power control illustrating how the (SIR) set command is executed, along with the related graph depicting the changing SIR as the UE continues to physically move, is shown in Figure 8.8. The base station tags each of the users' data frame with a frame reliability indicator, a process that can be thought of as a cyclic redundancy check (CRC), which is acquired during the decoding process for the particular user data frame in question. If the frame quality indicator reports back to

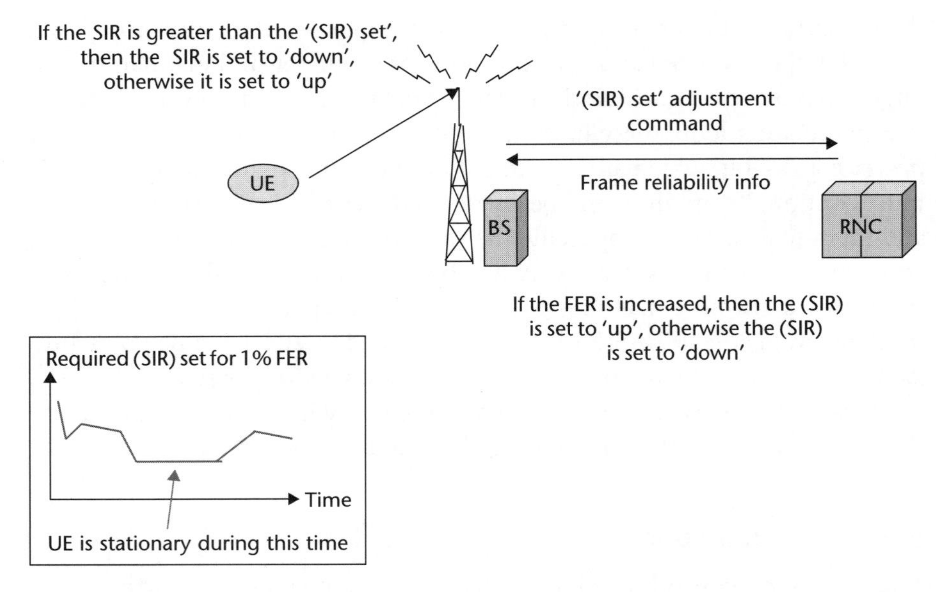

Figure 8.8 Outer loop.

the RNC that a decrease in transmission quality is occurring, the RNC will then command the base station to increase the target '(SIR) set' point to compensate. The outer loop mechanism is designed to be located in the RNC to ensure that accurate power control can also be maintained after any combined soft handovers have occurred.

8.6.2 Fast Power Control

Fast power control is an obstacle to soft handovers. Therefore, two factors need to be considered that differ from a single link connection: the power drifting of the base station downlink power and the detection of the uplink power commands in the UE, as illustrated in Figure 8.9. The problem occurs when a single command is sent to control the downlink transmission power by the UE, as this is received by all the base stations currently in the active set. This is due to the fact that the base stations will detect the received command from the UE independently, as the power control commands are unable to be combined in the RNC, therefore the signalling load will increase phenomenally. This will cause unacceptable

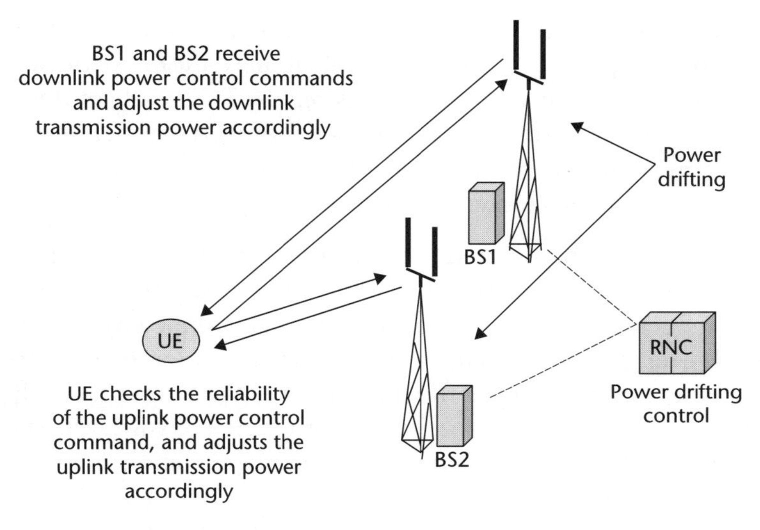

Figure 8.9 Power drifting.

delays, thus reducing the power control accuracy. In such a case, errors caused in the signalling can cause the base stations to incorrectly detect and process the control commands. This could lead to one base station increasing its download transmission power, while simultaneously the other base stations present in the active set decrease their downlink transmission power. Hence this is known as power drifting, as the downlink power starts drifting apart. This is counter-productive and causes the downlink soft handover performance to suffer. However, it is possible to control this from the RNC by introducing certain limits for the downlink power control dynamics. These power limit settings should be relatively strict and refer to the specific transmission powers of the UE. The goal is to minimize the UE's permitted power control dynamics as much as possible, as this in turn will reduce the maximum power drifting. However, power control performance is generally improved by large power control dynamics. Alternatively, information from the base stations can be received by the RNC concerning the transmission power levels of the currently active soft handover connections. An average level is then computed from numerous power commands, for example, say, over a 500 ms period. The RNC is then able to send a reference value based on this average level for the downlink transmission power to the base stations. Power drifting can be reduced by the base stations, currently in the soft handover scenario, by using this reference to compute their downlink power control. It is then possible to periodically perform a small correction to the reference power level as required. The size of this correction is directly proportional to the relationship between the reference power and the present actual transmission power, which will therefore minimize the amount of power drifting.

Fast fading can be compensated for by fast power control, and in an optimal scenario, will eliminate the fading experienced in the receiver, therefore reducing the required Eb/No. However, compensating for the fading will cause peaks to occur in the UE's transmission power levels, thus affecting any intercell interference present. Apart from fast fading, the near–far effect can also be prevented, or restricted, with fast power control, as the UE power levels received in the base station can be equalized. However, due to the radio environment, it will not be possible to achieve optimal power control accuracy, as various errors will inevitably occur. These include incorrect step sizes, transmission and processing delays, and incorrectly decoded power commands. Additionally, the increased residual variation in the signal-to-noise ratio will diminish the system performance. Any performance improvement that can be achieved due to fast power control occurs at its maximum in a small delayed spread signal. With UE moving at slow physical speeds, the Eb/No is improved and thus the fast power control can compensate for any fading in a more efficient manner. A wide bandwidth system is able to achieve better diversity power control as opposed to narrow band systems, due to a limited multi-path diversity within the pico- and microcells and therefore fast power control is more effective than within macrocells.

8.6.3 Improved Quality with the Power Control Signalling

It is possible to reduce the UE transmission powers by using power offsets, which can be achieved if the quality of the power control signalling can be improved. In a scenario where a UE is currently engaged in a soft handover, a reduction in transmission powers can be made. This can be achieved by setting a higher power for the dedicated physical common control channel (DPCCH) than for the dedicated physical data channel (DPDCH) in the downlink. The power offset between the DPCCH and the DPDCH can be dissimilar for the different DPCCH fields, such as pilot bits, power control bits, and the transport format combination indicator (TFCI). An illustration of how the improved power control signalling quality can be achieved by the use of a power offset can be seen in Figure 8.10. From the figure we can see that there is a power offset

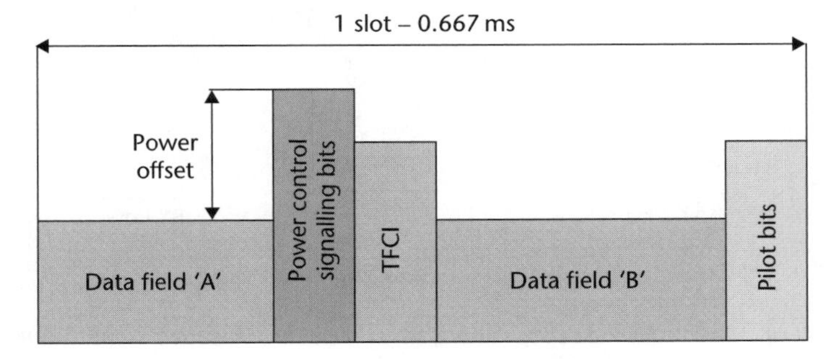

Figure 8.10 Improved power control signalling quality.

between the DPDCH and the DPCCH fields, hence this will in turn allow for a transmission power reduction within the UE. The pilot bits are used for the channel estimate in the receiver, while the power control signalling bits carry the power control commands for the downlink power control. The TFCI is transmitted in the DPCCH and its function is to inform the receiver which transport channels are active for the current frame along with carrying the rate information and indicating the transport format of the current frame. In this scenario it is possible to achieve reductions in UE transmission power of around 0.4–0.6 dB using power offsets.

8.6.4 Uplink Fast Power Control – Intercell Interference

Intercell interference will increase due to peaks in the transmission power of the UE. This occurs when the uplink power control is able to follow fast fading. Furthermore the adjacent cells will experience peaks within the interference received in the downlink, but due to the limited power control dynamics, the effect is reduced and is illustrated in Figure 8.11. The lesser the diversity available, the greater the transmission power, and thus more intercell interference will be generated. Systems utilizing fast power control that cannot exploit the multi-path diversity of the channel, and furthermore do not employ any antenna diversity, are prone to

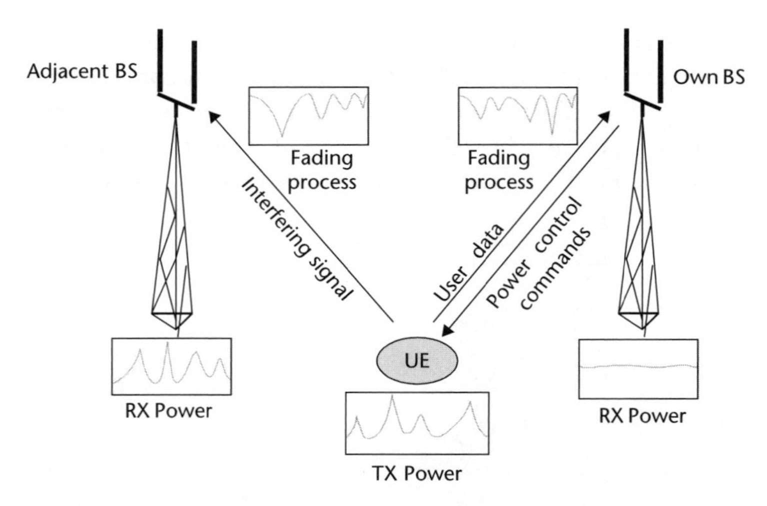

Figure 8.11 Interference to neighbouring base stations in fading channel with fast power control without macro diversity.

experiencing increased interference levels. Alternatively, soft handovers provide more diversity and can reduce intercell interference.

8.7 Handover Mechanism

When an active connection is in progress, the UE is continuously performing various measurements with respect to the adjacent cells and then reporting the status of these measurements to the RNC. These measurements are acquired from the adjacent cells' PICHs. Following this measurement, the RNC then evaluates whether the values indicated within the measurement reports will trigger any specific set criteria. If this is the case and they are triggered, a new base station will be added to the active set as shown in Figure 8.12. An 'active set' consists of a list of cells in which the UE is maintaining a connection to the network, or alternatively where a 'radio link setup' has been performed. Therefore, the UE may hold active connections between itself and the network with more than one cell simultaneously. At present, it is expected that an active set may consist of no more than three cells. It will of course be possible to increase the amount of cells further when required. However, although soft handover scenarios are advantageous, increasing the number of cells in an active set will in turn result in an increase both in signalling load and code consumption, the effect of which also needs to be considered.

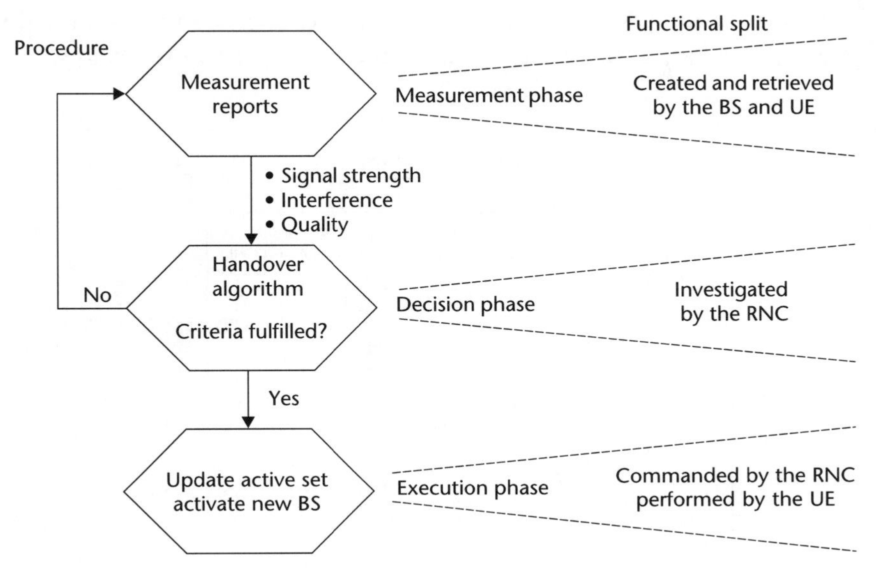

Figure 8.12 Handover mechanism.

Other parameters to consider are the candidate set, active set, neighbour set, remaining set, and discard set. The 'candidate set' comprises the base stations that fulfil the actual requirements to enter the 'active set', but are not actually included in it. The 'neighbour set' consists of the base stations that are possible candidates for a soft handover, and these are mostly the base stations whose coverage areas are in the vicinity of the UE. The 'remaining set' is the list of all remaining base stations not included in the other sets. Finally, there is the 'discard set' which contains the number of base stations that belong to the current active set, but are due to be discarded as they are no longer able to maintain the required parameters for the current active set. In addition, with regard to the handover mechanism, the following subsections cover the soft handover algorithm and handover measurements.

8.7.1 Handover Decision and Parameters

Soft handover performance impacts the required fade margin against shadowing and the number of users in a soft handover.

A third-generation UMTS (UMTS (3G)) handover decision is illustrated in Figure 8.13 showing the required adjustable parameters noted below.

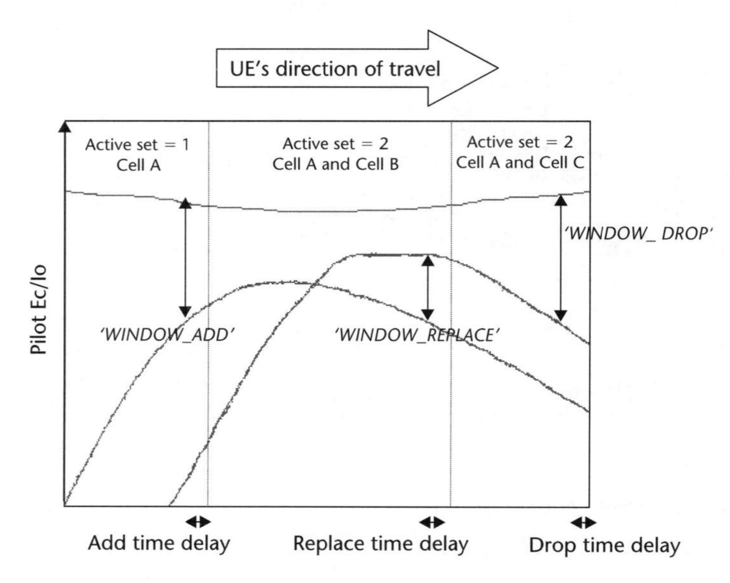

Figure 8.13 Handover decision and parameters.

Reducing the WINDOW_ADD parameter will reduce the number of handover channels required. Typically, networks will be aiming at having around 20–40 per cent of handover channels available. It should be noted that increasing the number of utilized handover channels reduces the network capacity. However, reducing the WINDOW_ADD beyond a certain point will reduce the effectiveness of a soft handover and lead to a reduction in macro diversity gain and an increased probability of handover failure.

The WINDOW_DROP and WINDOW_REPLACE are used to prevent a 'ping-pong' effect (hysteresis), causing cells to be repeatedly added and removed from the active set. As such, the WINDOW_DROP must be larger than WINDOW_ADD, including an added margin. However, reducing WINDOW_DROP will again reduce the number of UEs in a soft handover scenario. Description of handover parameters showing typical values are indicated in Table 8.1.

Handover parameters	Standard values	Parameter description
WINDOW_ADD	3 dB	This is the active set window, and every cell present in the active set window is added to the active set until the pre-defined level of AS_MAX_SIZE is reached
WINDOW_DROP	5 dB	Any cells present within the active set that fall below the set criteria or threshold are removed from the active set
WINDOW_REPLACE	3 dB	If the size of the active set is at AS_MAX_SIZE, then a current member is replaced when this threshold is breached
HCS_LAYER_UP/DOWN	Not applicable	The threshold levels required to select a higher or lower cell layer
HCS_LAYER_REPLACE	Not applicable	The thresholds required to replace the current layer (frequency) used with a cell layer
PS_MINIMUM	Not applicable	This is the minimum pilot strength of an active set pilot
AS_MAX_SIZE	Not applicable	This is the maximum size of the active set

Table 8.1 Handover parameters.

8.7.2 Soft Handover Algorithm

The purpose of this subsection is to give the reader an understanding of the basis of a soft handover algorithm showing the parameters involved and does not delve into the finer elements of building such an algorithm. The illustration in Figure 8.14 is an example of the W-CDMA soft handover algorithm. The pilot channel Ec/Io is used for the handover measurement quality which is signalled to the RNC via layer three signals. Further to this the following parameters are also shown in Figure 8.14:

a) 'Reporting range' is the threshold for soft handover.
b) 'Hysteresis Event 1A' is the additional hysteresis.
c) 'Hysteresis Event 1B' is the removal hysteresis.
d) 'Hysteresis Event 1C' is the replacement hysteresis.
e) ΔT is the time to trigger.

Hysteresis is used mainly to prevent ping-pong type handoffs; it can also be beneficial for traffic steering. A useful default value is considered to be 4 dB. With a hysteresis of 4 dB, a handoff will not occur from Cell A to Cell B, unless Cell B is detecting the UE more effectively than Cell A by a margin of 4 dB. This principle can also be applied regarding the downlink received signal strength indication.

8.7.3 Handover Measurements

As mentioned in the previous subsection, the pilot Ec/Io measurements are considered important criteria with regard to the accuracy of the handover measurements, which can also be seen in the previous illustration in Figure 8.14 (soft handover algorithm).

However, the optimal target which can be achieved with regard to handover measurements is one where all fast fading can be completely averaged out. To average out any fast fading present, a sample of each 10 ms frame can be taken and thus an average of the fast fading can be calculated. However, this average calculation can be impeded due to the effect of filtering, which can cause large substantial errors, as the filtering is performed at a rate of 100 ms. Hence, attempting to filter any fast fading within such a short period is not possible. This in turn is likely to cause

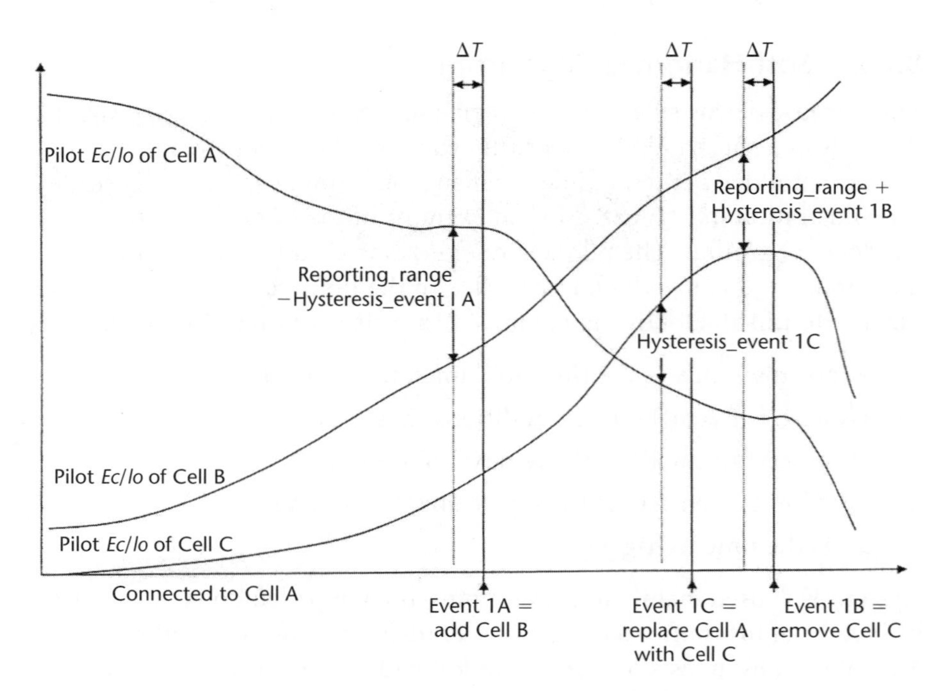

Figure 8.14 Soft handover algorithm. Source: WCDMA for UMTS, *Holma and Toskala.*

unnecessary handovers to occur due to measurement errors, such as those encountered by the filtering inadequacies. This will lead to an increased volume of handover signalling and shorter active set update periods. It is possible to improve the measurement accuracy by increasing the filtering length to 1 second. Long filtering periods are beneficial for UE's that are travelling at low physical speeds, as greater measurement accuracy can be achieved. At physical speeds of around 50 km/h, a filtering period of 100 ms will ensure that sufficient performance can be achieved and thus only minor improvements are likely to be gained by employing a longer filtering period. Although long filtering periods can be advantageous in certain circumstances, under other conditions they can cause delays to be induced, thus slowing down the handover process. This is particularly prevalent in the case of fast moving vehicular UEs, whereby pathloss conditions can experience rapid changes, for example, when moving around corners. An example of a fast moving UE currently employing a high bit rate service can be seen in Figure 8.15, depicting the effects of

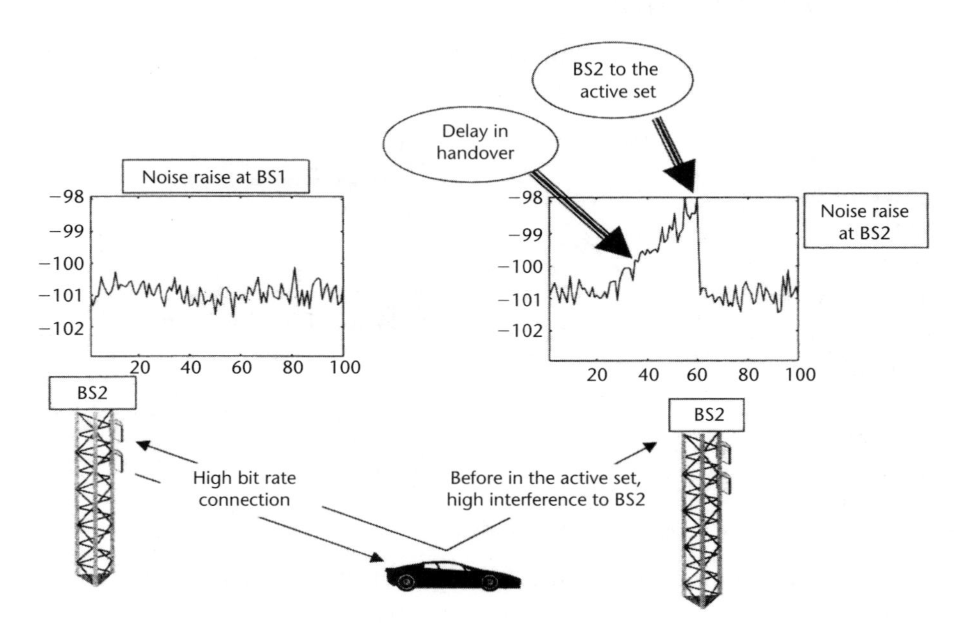

Figure 8.15 Handover measurements. Source: WCDMA for UMTS, *Holma and Toskala.*

a delayed handover. All the time base station two (BS2) is not present within the active set of the UE, it is unable to control the uplink transmission power and thus noise rise peaks can be caused at BS2. This is only likely to occur if there are long delays in the handover occurring due to excessive averaging in the measurements, delays in the handover signalling, or if the UE is utilizing a high data rate service and moving at high speed. Therefore, it is not considered advantageous for long filtering periods to be used in the handover measurements, as the ideal filtering period consists of a trade-off between handover delay and measurement accuracy.

8.8 Multi-path Components

There are two multi-path components to be considered: macro- and micro diversity which will enable better call quality to be achieved.

8.8.1 Macro Diversity

With macro diversity, the functionality must also exist at the RNC level as it is likely that a UE will hold connections to different base stations, possibly even through a base station parented to a different RNC. An example is illustrated in Figure 8.16 depicting a UE holding three connections simultaneously. This means that a three-cell active set is in use, and one base station is connected to a different RNC. In such a case, the base stations perform the signal summing within their RAKE receivers first for the radio paths of their own, and then the end result consists of the final summing being performed at the RNC level. With multi-path signals, it is possible to re-construct the multi-path

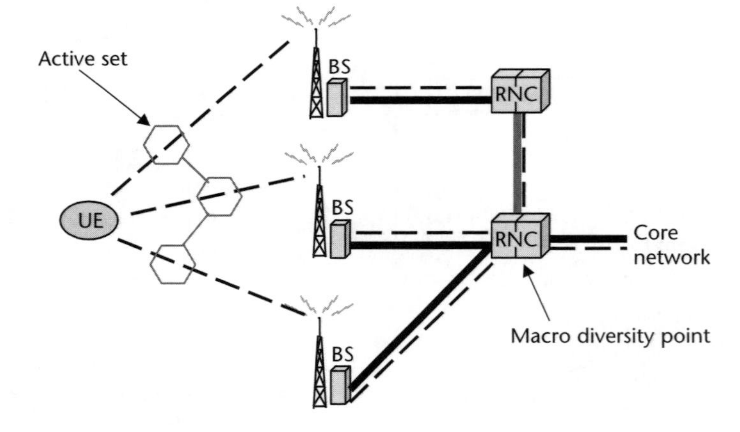

Figure 8.16 Macro diversity.

components into a reinforced signal, thus resulting in an improved signal quality. As described earlier in this chapter regarding power control, it is counter-productive within a W-CDMA system for the UE to transmit with power levels that are too high, as this will increase the interference levels in the cell, reduce the cell size, and is likely to start blocking to occur. The optimal method to increase call quality is to use multi-path propagation. Concerning soft and softer handovers, one of the negative aspects is the extra radio capacity consumed when holding more than one connection. However, the system capacity is increased due to the added capacity gained from the reduction in interference.

8.8.2 Micro Diversity

Micro diversity can be defined as the propagating multi-path components that are combined in the base station. This is where the RAKE receiver in the base station is able to determine, differentiate, and combine numerous received signals in the radio path as illustrated in Figure 8.17. As the radio frequency (RF) propagates, the signals are reflected and refracted due to the RF 'hostile' environment, and can be received at the base station with a

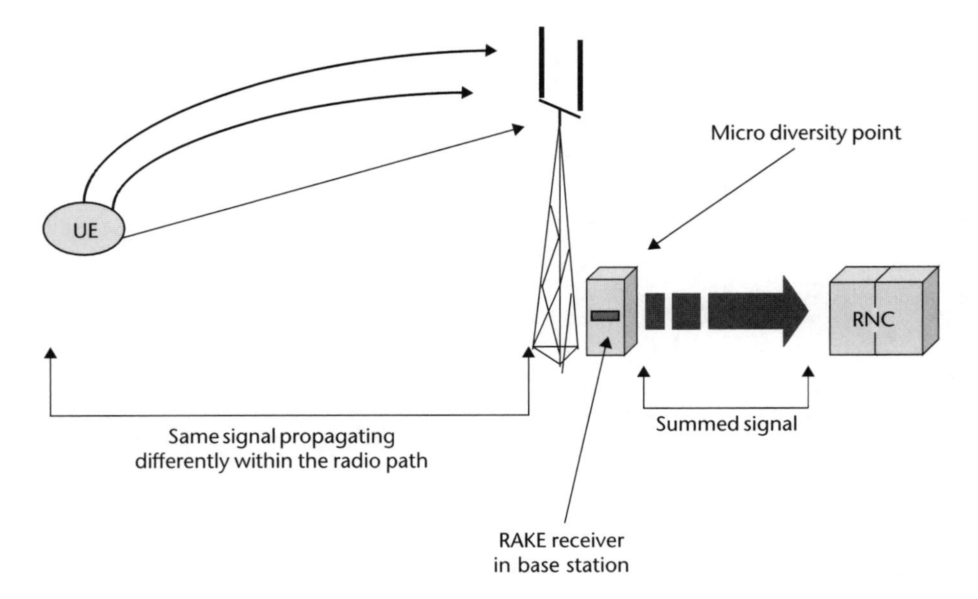

Figure 8.17 Micro diversity.

reduced amplitude, as well as being out of phase. So the micro diversity functionality is able to combine the different signal paths received from one cell, and with regard to a sectored base station is able to combine the signal paths from different sectors. This scenario would be the situation experienced in a soft handover.

In summary, RRM functionalities ultimately provide the optimal results for the three CCQ parameters, therefore guaranteeing a certain target QoS, maintaining the planned coverage area, and ensuring the capacity requirements can be fulfilled. This chapter has covered the important RRM functions which will have an impact on the overall system efficiency and in addition, the infrastructure costs.

Summary for Part IV

This concludes Part IV with regard to the third parameter of the CCQ model, QoS. The major issues covered in this part have been the QoS classes, service capabilities, and the mechanisms available. In addition, RRM issues which all have a direct effect on quality have been discussed. The admission control, load control, packet scheduler and power control must all be taken into consideration along with the handover mechanism. It is important that each of these parameters is understood, as this will in turn enable the planner to correctly construct the network plan while ensuring the required QoS can be achieved at the optimal levels possible.

This leads into the next part of the model, optimization and network Planning.

PART FIVE

Optimization and Network Planning

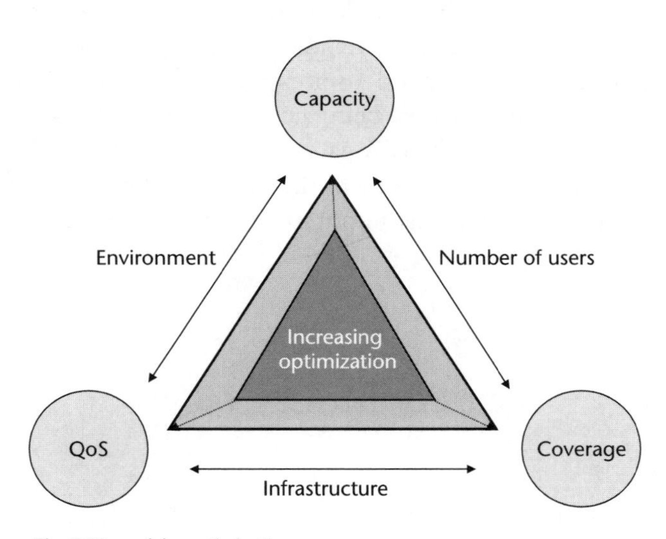

The CCQ model – optimization.

Introduction

Optimization can be considered as the final and on-going process within a third-generation universal mobile telephony system (UMTS (3G)) network and whose function is to ensure that the optimal coverage, capacity, and quality-of-service (QoS), abbreviated as CCQ, levels can, and will, continue to be provided. It is likely that in the beginning the network operation optimization engineers will have access to key data from the radio access network (RAN) elements. This information will be essential in performing the initial network optimization.

Within this section, planning methodologies, carrier wave measurements, radio resource management (including admission and load control), packet scheduling, inefficient sites, and antenna heights are just some of the subjects covered with regard to network optimization.

Once the network's operational stability is ensured from regular network monitoring, which could be at 15 min intervals or less depending on when an identifiable pattern of traffic and data usage emerges, will allow the optimization engineers time to compensate accordingly. The interaction that will occur between the traffic, RAN, and the services being used, could become somewhat problematic as there will be very limited experience on their effect on the CCQ factors. The goal therefore is to effectively manage the 'end to end' user experience either in terms of his or her data requirement, the area being covered or the effects of not being able to achieve the optimal QoS for each and every user regardless of the service being used. This will be one of the major challenges facing any 3G operator and emphasizes the importance of optimization over the long term.

CHAPTER NINE

Radio Environments and Microcell Planning

This chapter covers the issues with regard to microcell planning in the network design, taking into account the soft handover design elements, indoor to outdoor environments, and vehicular environments. As the bulk of revenues likely to be generated will be from high data rate users located in densely populated areas, this will require accurate microcell and picocell planning. Due to the high volume of small cells required, it is critical for the reader to understand the inherent differences and associated problems likely to occur with this type of planning as this will have an effect on all parameters within the coverage, capacity, and quality-of-service (QoS), abbreviated as CCQ.

9.1 Microcell Planning

With a microcell, the radio propagation environment differs significantly from a macrocell, and hence as with all cell planning, microcell planning must be accurately performed. A macrocell can be considered as a cell having a radius of 1 km or more. A microcell would cover a relatively small area with a radius of a few hundred metres, and a picocell would typically be used to provide coverage in an indoor area covering around 50 m to around 100 m. Figures 9.1 and 9.2 indicate the typical affected areas within the network plan that benefit from microcells. One of the characteristics with small cells such as microcells is that they will generally be required in areas that consist of high amounts of clutter,

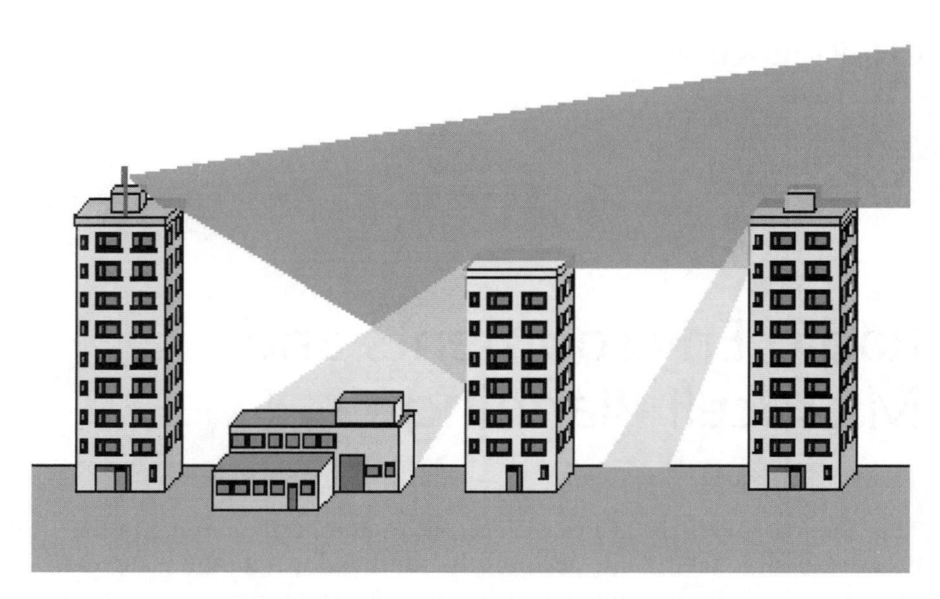

Figure 9.1 Microcell planning (1). Signal reflection can be used to cover specific areas.

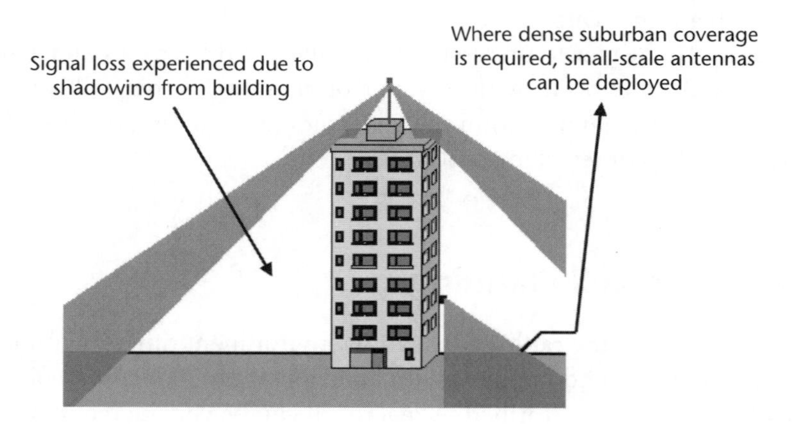

Figure 9.2 Microcell planning (2).

for example inner city streets. In this type of environment it would be advantageous to install the base stations at the same height as street lamp-posts. This will result in the radio frequency (RF) signals propagating along the street canyons. With this in mind, the two most important factors that must be considered here are soft handover design and corner effect.

9.1.1 Corner Effect

In a dense urban city environment the received signal level in the user equipment (UE) will experience rapid changes. The UE will experience too fast a rise in signal strength when it turns the corner if there is another base station located around that corner. This is likely to create problems, for example, if the UE is unable to 'lock on' to the station fast enough, the interference increase experienced can cause the call, or transfer, to be dropped. In addition, as the transmission power of the UE is not yet under the control of the new base station, its high power level will be seen by the new base station and consequently, the existing users connected to this new base station will be forced to increase their own transmission power to the same level. This will cause more interference to be generated and ultimately lead to more dropped calls. This problem is known as corner effect (illustrated in Figure 9.3) and can be reduced if it is possible to execute a fast forward handover, simultaneously dropping the old base station.

A fast forward handover decentralizes the handoff decision, as in the case of a UE-assisted handoff, which is often able to achieve the fast forward handoff decision. Decentralizing the handoff process is beneficial as the core network does not get involved in making the handoff decisions for every user, and therefore this also reduces the signalling load.

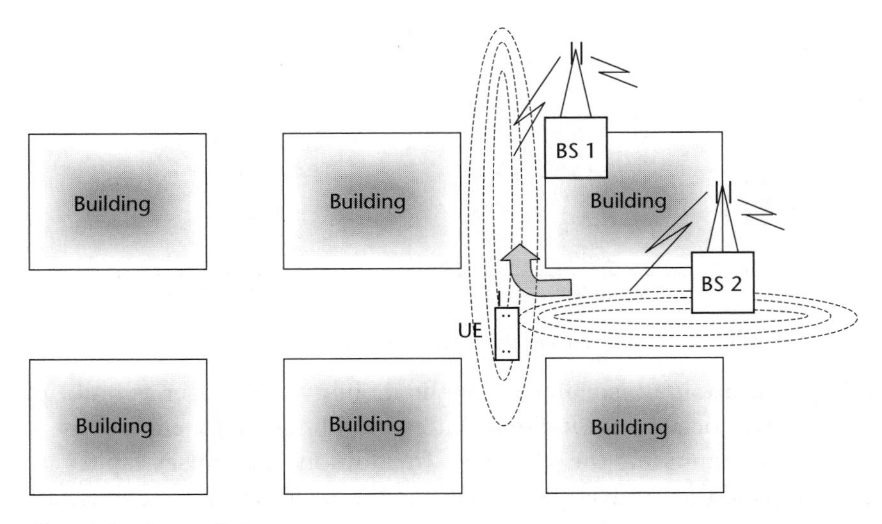

Figure 9.3 Corner effect.

Correct cell location planning, defining handover thresholds and planning cells that overlap each other can resolve the problems encountered by corner effect. Overlapping cells are considered advantageous, as they can allow the UE to enter into a soft handover state just before entering and leaving a street corner. This can be achieved by defining the handover threshold parameters, covered in Section 8.7.

9.1.2 Pathloss Attenuation

Pathloss attenuation describes the RF signal strength losses that exist between a separate transmitter and receiver. Mostly RF losses occur due to the presence of clutter, however, losses also occur within free space and are referred to as the free space pathloss. In general, the greater the distance between the transmitter and the receiver (TRX), the greater will be the RF pathloss. Fairly large pathloss attenuation can occur as can be seen in Figure 9.4 with respect to both line-of-site (LOS) and no line-of-site (NLOS), caused by numerous obstructions along the RF path or link, such as trees, buildings, uneven terrain, etc. A variation of around 10–12 dB is considered to be the standard deviation figure assumed for obstructions or shadowing. However, when considering average building penetration losses, a standard deviation of around 8 dB should be assumed.

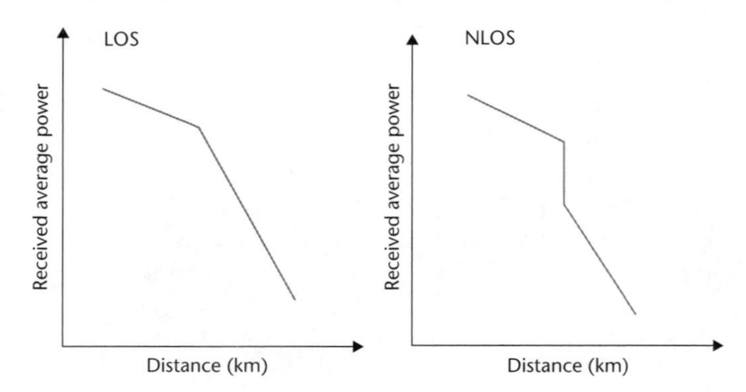

Figure 9.4 Microcell propagation in LOS and NLOS situations.

Radio propagation is also affected by the mobility of the user as multi-path scattering will occur from various clutter (buildings, trees, etc.), thus resulting in attenuating the RF signal. In addition, this causes rapid fluctuations of the received RF signal known as fading. A basic illustration can be seen in Figure 9.5.

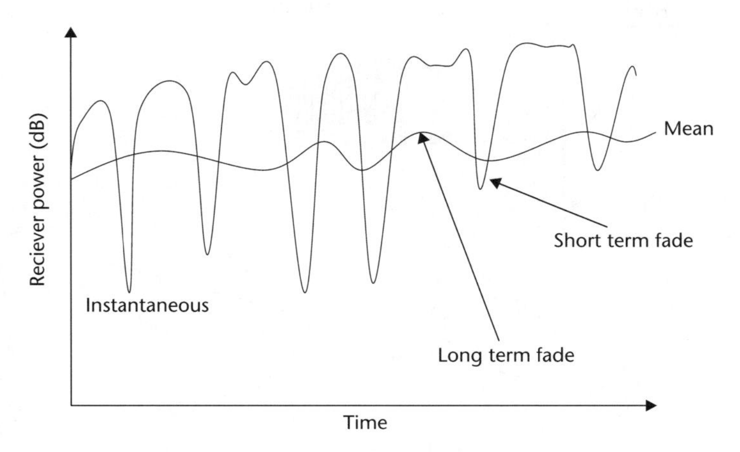

Figure 9.5 Radio frequency propagation.

Large-scale or long-term fades occur within free space, causing signal attenuation and therefore power degradation.

Small-scale or short-term fades occur due to rapid changes in signal strength over a small area or time interval. The received signals may also suffer fading due to movement of surrounding objects such as vehicles.

With regard to small-scale fading, low delay spreads will occur (approximately 0.2 μs) and will consist of either Rican or Rayleigh fading. Rayleigh fading occurs where multiple indirect paths exist between the TRX with no distinct dominant path, therefore no clear desired signal is present. Hence, with a Rayleigh distribution, a sum of multiple, independent variable signals arrive at the receiver. Rayleigh distribution is a deep fading process characteristic of radio signals within a multi-path propagation environment. An illustration of multi-path or Rayleigh fading can be seen in Figure 9.6. Rican fading is the statistical energy distribution of a direct wave path from the transmitter to the receiver. It is also referred to as the LOS path and represents the variation in signal strength that occurs when the path from a transmitter to a receiver is not obstructed; this situation can be caused by atmospheric conditions. Small-scale fading can be considered as a more rapid fluctuation of signals usually caused by multi-path signals, Doppler effects, and signals that may be wider than the coherence bandwidth of the radio channel in use. Rican fading describes a condition that occurs when one dominant signal arrives at the receiver

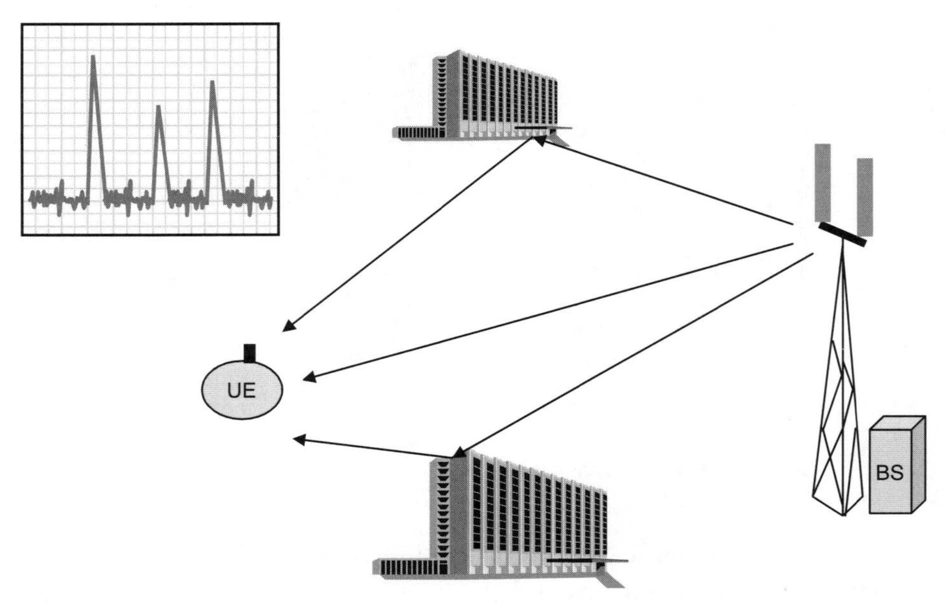

Figure 9.6 Multi-path (Rayleigh) fading.

with several other weaker multi-path signals. Rican fading is not common in two-way communication as buildings or other objects usually obstruct an LOS to the source.

9.1.3 Soft Handover Design in Microcell Environments

As users travel through the streets, numerous handovers will frequently occur within microcells, and they will experience corner effect as previously described. If the soft handovers being executed are occurring too slowly, this is likely to generate large amounts of interference, therefore decreasing the capacity of the cell. The planner must therefore be aware that standard cell planning is not advisable for microcell environments. In addition, where there is a high volume of microcells, there will be a high volume of soft handovers in progress, hence this will cause the signalling load to increase dramatically. The recommended planning strategy is to attempt to force a reduction in the number of soft handovers. This can be achieved by using distributed antennas and sectored cells. Sectorizing cells consist of dividing the existing cell into sectors. For example, an omni-site comprising say, one TRX, could be split into three sectors, thus employing a separate TRX for each sector, as illustrated in Figure 9.7. The use of sectors within the cell will also have the effect of increasing the capacity of the cell. Distributed antennas are antennas that are distributed within a defined space, as illustrated in Figure 9.8. By using distributed antennas, the volume of micro and picocells and therefore the cost of implementation can be significantly reduced. The downside here is that considering the UE is only in one location, the forward link capacity is not as desirable, as the signal is transmitted to the entire coverage area regardless of the UE's location.

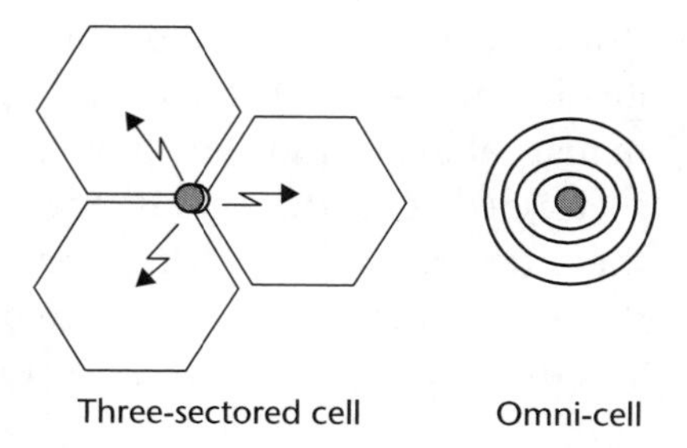

Figure 9.7 Sectored cell.

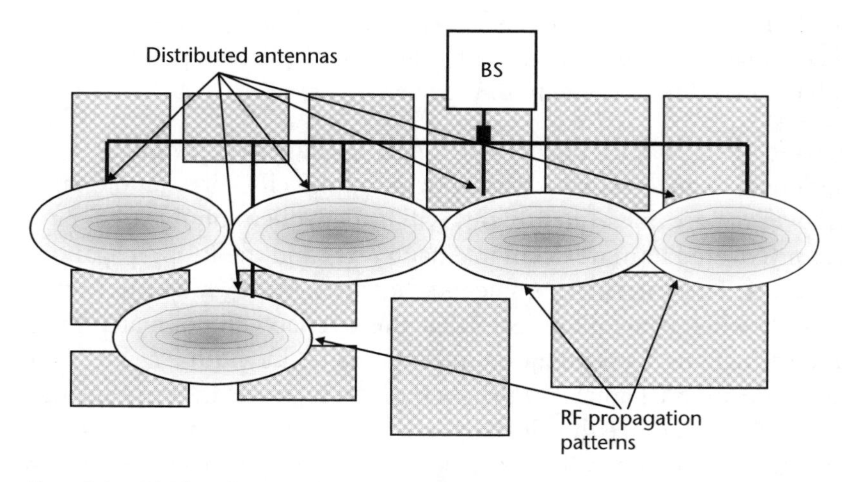

Figure 9.8 Distributed antennas – street microcell.

In summary, the main advantage of employing the distributed antenna method is that a reduction in the number of base stations can be made resulting in significant cost savings; a secondary benefit is the reduction of the system's signalling load. The disadvantages are mainly implementation as well as planning issues, due to the requirement of installing large antenna cabling and the consequential signal losses that will occur throughout this cabling. In addition, the downlink RF signal must be more carefully controlled, as it is likely to propagate further than required.

With sectorization, cell splitting can be advantageous in reducing cell loading and spreading the load throughout the coverage area in a more manageable fashion. The downside here is with an increased number of sectorized cells; this will increase the softer handovers and in turn increase the interference levels, thus causing a reduction in capacity. Furthermore, this will also cause trade-offs to occur with regard to both sectorization and distributed antennas, so a suitable balance must be achieved.

Within universal mobile telephony system (UMTS) it will be possible to have six-sectored sites, which in turn will enhance both coverage and capacity. The increase experienced in the RF forward gain when using six narrow beam antennas will in turn cause an increase in the coverage area. In addition, the capacity increase will be doubled when compared to that of a three-sectored cell. This increase in antenna gain, can be attributed to the narrow beam width of the antenna, however the gain is likely to be only 2 or 3 dB. Although this is beneficial, it will not have a great effect on the coverage area, however if sectorization is required it will ultimately be implemented due to capacity limited areas.

A six-sector site can lead to an increase in coverage, as previously mentioned, in an area that is served by multiple cells, for example the soft handover region. Subsequently this will be dependent on the antenna pattern (covered in more detail in Chapter 11) and the RF propagation conditions. This can cause problems whereby signal overlap can occur, especially when using more sectors, as this overlap will not match the soft handover regions. Dependent on the antenna beam widths (again covered in more detail in Chapter 11) the overlap can increase, in turn causing unwanted interference thus leading to a reduction in capacity. This unwanted interference will be reduced by the soft handover mechanism, which can utilize the overlapping area between the sectors as a form of macro diversity. However, it should be noted that due to this reduced unwanted interference, even despite any soft handovers, a reduction in downlink capacity will be experienced. Thus it is unlikely that a doubling in capacity would be experienced by increasing from three to six sectors.

In a real-life scenario it is likely that the amount of overlap will be greater due to the effect of the adjacent sites. In addition, the planner should be aware that in this case the use of sectorization will require accurate radio planning to ensure any coverage holes between the cells are reduced, while simultaneously minimizing the overlap of any neighbouring cells.

With regard to the soft handover design, to ensure they are effective, the UE receivers must contain sufficient RAKE fingers (RAKE receivers are covered in Section 3.7). The greater the number of overlapping sectors, the more RAKE fingers will be required. In addition, having more sectors will place a greater load on the radio network controller (RNC) and finally the deployment issues of implementing six antennas must also be considered, especially where limited space is available in densely populated urban environments.

9.2 Radio Environments

9.2.1 Indoor/Office Environment Planning

Low levels of RF transmit power from both the UEs and the base stations will be required in indoor environments. Likely losses will be in the ranges of around 15–20 dB for solid concrete walls, to around 3 dB for lighter materials as illustrated in Figure 9.9. This will vary depending on the construction materials, however, a standard deviation of 12 dB is assumed for shadowing. When the UE is present in the same room with the base station Rican fading will occur. Alternatively, Rayleigh fading will occur when the UE is located in a separate room or possibly on a separate floor from the base station. Therefore, both Rican and Rayleigh fading are likely to occur within the indoor environment. The coherence bandwidth here will be rather large and thus the diversity order or path diversity will be somewhat reduced due to the short propagation distances involved.

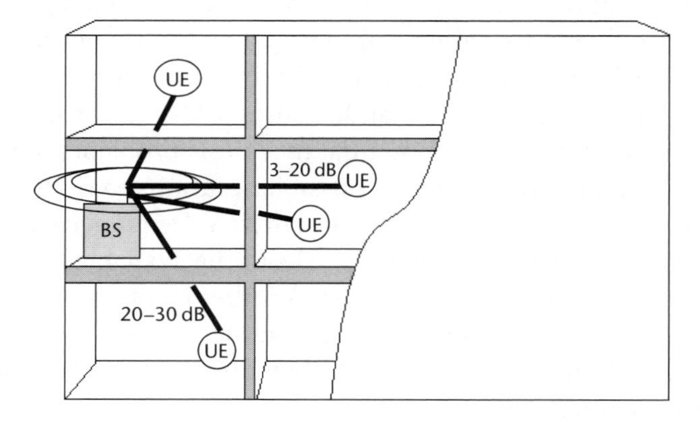

Figure 9.9 Varying attenuation in indoor/office environment.

Coherence bandwidth can be thought of as the approximate maximum bandwidth or frequency interval over which two frequencies of a signal are likely to experience comparable or correlated amplitude fading. It must also be taken into consideration that coherence bandwidth varies over radio communications paths due to the multi-path spread varying from path to path.

If the data rate is slow, such that the modulation bandwidth is less than the coherence bandwidth, frequency selective fading can be avoided. Nevertheless, the received signal will be subject to flat fading. With flat fading the channel frequency response is identical across the bandwidth of the transmitted signal.

Any delay spreads occurring will be minimal within this type of in-building environment. Improvements would be noticeable with the received delayed paths due to the operation of the RAKE receiver. One solution here could be to deploy distributed antennas, however, this would increase the multi-path RF products received by both the UE and the base station. Careful planning will be required when both indoor and outdoor cells are using the same frequency. One challenge to be overcome will be to ensure the soft handoff parameters are carefully planned, as all UEs must 'connect' to a cell where the minimum amount of transmission power is used. It will certainly be possible for outdoor users to occasionally 'acquire' picocells through windows. One option to reduce the possibilities of this occurring would be to create Hierarchical Cell Structures (HCS) (see Figure 3.6) which would also reduce the chances of UEs adding picocells into their active sets. This would ensure that a macro or microcell would be overlaid on top of the aforementioned picocell, thus allowing the UE to add the larger cell to its active set. However, it is unlikely that a UE located outside will hold this connection for very long, so it will be critical to ensure the handoff thresholds are correctly specified, thus minimizing any potential hysteresis problems (covered in more detail in Chapter 8). There are possible solutions to resolve these issues, such as deploying indoor cells on different frequencies (assuming the operator has purchased adequate blocks of the available frequency spectrum, covered further in Chapter 5). Another solution would be to strategically locate the antennas and position them where they are not directed towards windows, simultaneously reducing the indoor base station transmission powers.

9.2.2 Outdoor to Indoor Pedestrian Environment

When planning in an outdoor to indoor pedestrian-type environment, positioning analysis with a Manhattan model, as illustrated in Figure 9.10 is recommended. An outdoor to indoor pedestrian environment refers to providing both outdoor coverage and indoor coverage from outdoor sites. The number of microcells will need to be determined and will be dependent on the coverage required. However, the key is to control the RF emissions, thus keeping the transmit powers to a minimum. This factor should always be kept in mind and can be considered one of the key issues with regard to UMTS planning. When considering a Manhattan model, microcells can be both deployed in the middle of street canyons and also on street corners.

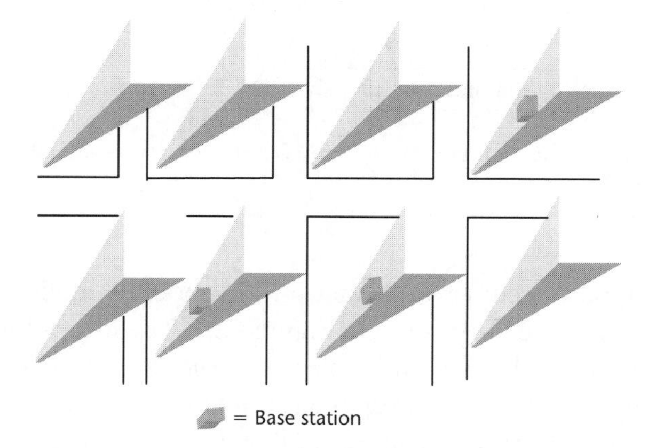

Figure 9.10 Manhattan microcell deployment model.

9.2.2.1 Manhattan Example

To perform a simulation in a street environment, the coverage area required consists of equally sized roads or streets, along with identical clutter (e.g. for RF propagation simulation purposes all buildings are assumed to be the same size). The model ensures that all possible coordinates, where a UE is 'positioned' have an equal chance of establishing a network connection. The UEs will move along the streets at a mean speed of 3–4 km/h, therefore the probability of turning at any one crossing (turning probability (TP)) is shown in Figure 9.11. Additionally the speed that the UE is moving can be varied, thus a change can be initiated at each position update corresponding to a given probability of change.

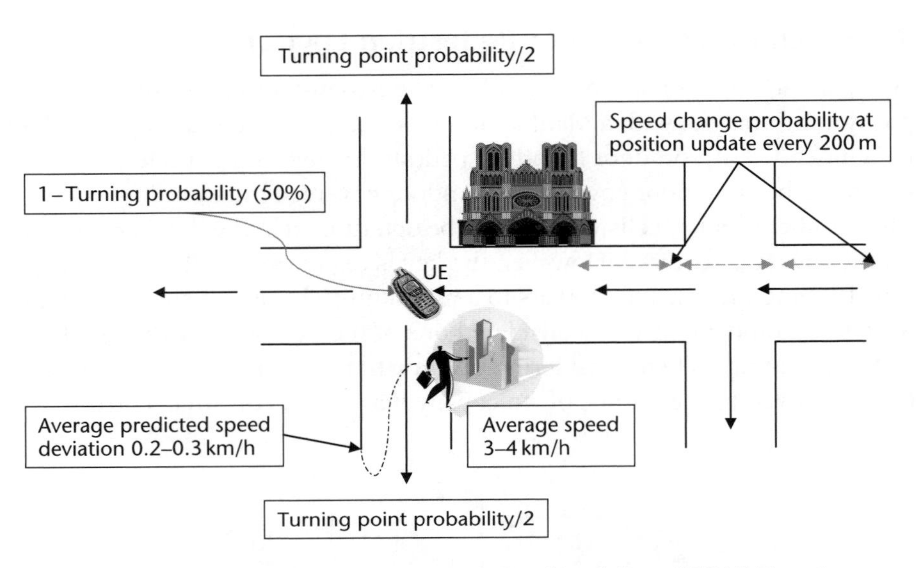

Figure 9.11 Turning probability. Reproduced by permission of Artech House: WCDMA Towards IP Mobility and Mobile Internet, *Ojanpera & Prasad.*

This position update occurs every 5 metres. The mobility model parameters are also indicated in the illustration in Figure 9.11. The idea is for the direction of each UE to be randomly selected at the start of the network connection. The UEs that exit the defined simulation area are naturally lost and the density of the traffic becomes variable. So, an incoming probability of new UE's entering the coverage area must equal the UE's leaving probability, to ensure traffic density consistency with regard to the periphery of the entire simulated area. This is accomplished by a simple procedure of creating a new user immediately after another user departs the simulation area.

9.2.3 Vehicular Radio Environment

A vehicular environment will use large macrocells radiating high RF transmit powers. To explain this scenario the macrocells are modelled in a consistent manner of a hexagonal framework as illustrated in Figure 9.12. The majority of received signals will consist of reflections. In this situation, standard deviation varies significantly. For example, a general assumption can be made of around 10 dB, which can be used when dealing with urban and suburban areas. As the received power at high

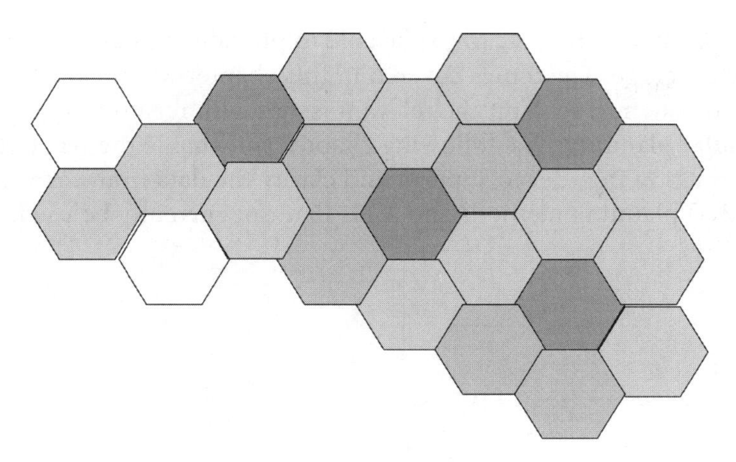

Figure 9.12 Layout of macrocells.

speeds will also vary considerably, this can be modelled with a log-normal distribution. In mountainous and rural areas lower values for standard deviation can be applied. The mobility model recommended is a pseudo-random model, which has semi-directed trajectories, as it must be able to compensate for the fact that the UEs will continually move within the entire simulation area. The directed trajectories in this case are used to generate a set of abstract assertions including implementation mapping for the required simulations. The simulation patterns for analysis are then mapped against a trajectory graph. Since the trajectory graph maps out complex interactions, it can be used as a coverage metric to quantify the set of simulation patterns required for analysis. In addition, the UE's directions are selected in a random manner after initialization, and will be uniformly positioned over the simulation area. It is possible to construct a simple vehicular mobility model, whereby a position update will occur after every de-correlation length of 15 to 20 metres. The de-correlation rate refers to the signal orthogonality as compared to other users. Hence it can be assumed that for a user traveling at around 100 km/hr, a direction change and update is likely to be between 40° and 45°. It should be noted that the update is independent of the previous direction update.

To summarize, as UMTS (3G) will offer a layered approach to providing coverage, small densely populated areas such as office environments, business parks, airports, etc. must be able to provide high quality, high data rate coverage. The topics covered in this chapter will enable the planner to take into account the relevant issues with regard to micro and pico cellular planning. The following chapter will enable the reader to use this information in order to support and clarify the data from the various UMTS (3G) planning methods and tools that can currently be used.

3G Planning Methodologies and Tools

This chapter covers the type of planning and optimization methods which will enable the planner to determine which types of radio frequency (RF) propagation models are best suited to the variety of different and challenging environments ranging from densely populated urban and inner city areas to suburban areas.

Also reviewed are the fundamental aspect of third-generation (3G) planning, which is represented by the Monte Carlo simulations and the various parameters required. These methodologies have a distinct parallel relationship with the coverage, capacity, and quality-of-service (QoS) model, abbreviated as CCQ model, as this is the basis of the Monte Carlo simulation, which is to provide the most efficient combination of CCQ and, therefore, is the ideal optimization tool. Finally, the importance of the mapping data and the test measurements for calibration purposes are also explained.

10.1 Planning Methodologies

Within 3G universal mobile telephony system (UMTS (3G)), three types of planning methods can be considered: static analysis (calculations), dynamic simulations, and static simulations, the static calculation method being a similar system to that used in global system for mobile communications (GSM) planning. This utilizes a deterministic algorithm within the planning tool, and thus a statistical analysis of the network is used to derive the following design thresholds:

a) Soft handover gain (typically 5 dB at the cell periphery)
b) Interference margins (both intracell and intercell)
c) Fade margins (to design a given coverage probability)
d) Special technique margins (adaptive antennas, transmit diversity, etc.).

Using the simulation method, it is possible to break this down into the two types that could be utilized: static simulations and dynamic simulations. Static simulations analyse the performance of a 'snapshot' of the network. This is an instance in time with user equipments (UEs) located in statistically determined places. Dynamic simulations simulate UEs moving through the network in successive 'timeslots'. The simulation may consider time to be split into chip periods, bit periods, and timeslots (signal-to-noise ratio (SNR) being considered). Successive 'timeslots' are then simulated taking into account the results of the previous 'timeslot'. New UEs are simulated both when entering into the network and terminating their connections. As each user is added into the simulation process, the major issues such as the interference levels created and the coverage are analysed. The whole iterative process is then repeated, adding another user, and so on. The various failure mechanisms are also considered, such as the UE's maximum power, the maximum base station power reached, the number of available channels, and the low-pilot Ec/Io. Once this process has been completed – usually around tens of thousands of 'snapshots' of an average 'view' of the network can be seen – the data can be interpreted. These mechanisms are the basis of the Monte Carlo simulation.

10.2 Planning Methodology Comparisons

It is likely that a combination of all three processes (static analysis, static simulations and dynamic simulations), illustrated in Table 10.1, will be utilized by an operator due to the different performance results of the available techniques, thus enabling both a fast network roll-out and efficient network design.

Static analysis would prove beneficial for the bulk of the planning, when site densities and actual locations are still at a preliminary stage, and static simulations could then be used for regional validation.

	Accuracy	Complexity	Time required
Static analysis	Poor – particularly with global margins (as seen from IS-95)	Relatively simple to use, once correctly configured	Fastest completion time, same as second generation
Static simulation	Fair – no dynamic network performance results possible	More complex to configure, more complex results	Reasonable, dependent on number of cells and UEs
Dynamic simulation	Likely to be high, assuming only minimal incorrect assumptions are made	Somewhat complex to interpret results	Long, especially if numerous runs are performed to give statistical validity

Table 10.1 Planning methodology comparisons.

For the final site selection, a relatively fast analysis of candidates is required. Confidence in final candidate selection must be high, so it would probably be beneficial to use static analysis to calculate an initial candidate shortlist. Static simulations could be used over a small area for the final selection of candidates, and also over a large area for final validation. For optimization, both static and dynamic simulations should be considered for high-sensitivity and problem areas (e.g. high-interference areas, street canyons, etc.) also detailed optimization, radio resource management, and call admission algorithms, such as admission control (covered in Chapter 8).

10.3 3G Planning Tool Utilization

There are various network planning tools available in the market and before considering a specific tool, the planner is recommended to consider the user's actual requirements and expectations from the tool in question. It may be useful to consider an integrated suite-type solution, which consists of standalone modules which may be used together to form an integrated tool set. This should encompass all areas of radio, network and transmission planning, simulation, optimization, statistical reporting, etc.

A typical tool 'suite' should consist of the following modules:

a) Radio planning
b) Test mobile analysis and auto-diagnostics
c) Cell site configuration design, equipment configuration, and inventory

d) Network performance monitoring
e) Reporting and management
f) Transmission, microwave link and backhaul planning
g) Carrier wave measurement logging tool
h) Network data and configuration management

Furthermore, some of the main functionalities required when considering the requirements of any 3G radio planning tool are as follows:

a) Radio coverage planning
b) Propagation modelling
c) Subscriber traffic planning
d) Network dimensioning
e) Antenna system engineering
f) Performed optimization of the implemented network
g) 3G service definition
h) 3G terminal-types definition
i) 3G cell parameters and carriers
j) Terminal density array creation
k) Monte Carlo simulator
l) Analysis reports
m) Results arrays
n) Batch runs
o) Dimensional planner
p) Code planner

The above-listed requirements have not been explained in detail as these should be thought of as merely a guide for the discerning planner or persons responsible for tool selection.

10.4 Planning Using Monte Carlo Simulations

Monte Carlo simulations should be considered as the key to 3G planning and will therefore ultimately provide the most accurate statistical predictions through the utilization of Monte Carlo algorithms, thus providing reliable and credible results. This is achieved by attempting to create many snapshots in time of the network using statistical distributions and many iterations, thus accounting for the randomness of the propagation environment (see Figure 10.1). This will result in deriving statistically valid

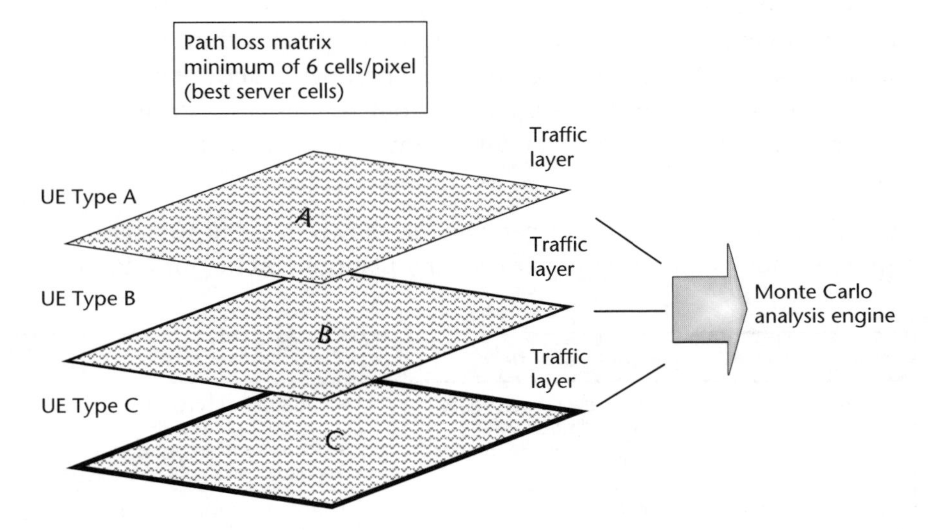

Figure 10.1 Monte Carlo simulations.

measurements of the likelihood of the network performance, for example how the network will perform with respect to CCQ. As there is a dynamic relationship between the network capacity and the network coverage in UMTS frequency division duplex (FDD), the location and amount of active UEs within the network must be taken into account, along with the transmission power of each active UE. It must also be noted that the transmission power of the UE will vary depending on the type of service being used. An effective method to model this relationship between capacity and coverage is to perform this iterative process using Monte Carlo simulations. The algorithm is capable of generating for a designated hourly period and geographical area, a raster (which can be thought of as a set of horizontal lines composed of pixels that are used to form an image on a video screen) of which pixels are most likely to represent a specific cell. Depending on the planning tool, it should be possible to define different user distributions and services to be included in one simulation.

10.4.1 Monte Carlo Input Parameters

The number of iterations required to achieve an accurate prediction, along with the resolution of the mapping data and propagation predictions can be defined and set. In addition, a boundary region for any 'edge effects' and the number of overlapping predictions can also be set within the prediction tool. Active terminals, which can be considered as terminals that are moving, are randomly allocated to each pixel within the simulation area. During each iteration, the number of active terminals represented by each pixel is Poisson distributed. Poisson distribution is used to model the number of random events occurring within a given time interval.

The following six sets of input parameters, shown in Table 10.2, are required before attempting to perform any planning predictions. Basic

Input parameters	Parameter criteria	
a) Simulation parameters	• BS receiver noise figure • Thermal noise value • Edge of cell size (metres)	• Number of serving (overlapping) cells • Number of snapshots • Slow fading standard deviation
b) Cell parameters	• Location • Propagation model (from radio planner)	• Antenna (from radio planner) • UMTS FDD carrier assigned the cell
c) Service parameters	• Required Eb/No or BER uplink • Acceptable Eb/No or BER uplink • Required Eb/No downlink • Acceptable Eb/No or BER downlink • Eb/No to BER look up table • Guaranteed uplink bit rate	• Acceptable uplink bit rate • Guaranteed downlink bit rate • Acceptable downlink bit rate • Support soft handover • Traffic switching type – circuit or packet
d) Terminal parameters	• Maximum TX power • Required Eb/No	• Power control step size (dB) • Dynamic range
e) Cell carrier assignment parameters	• Maximum users allowed • Maximum TX power • Noise limit • Pilot power • Number of primary channels	• Number of handover channels • Traffic channel orthogonality • Common channel power • Scrambling code
f) Circuit-switched services	• Number of packet calls/session • Size of packets in bytes • Reading time between packet calls	• Number of packets within a call • Interarrival time between packets • Re-transmission rate percentage

Table 10.2 Monte Carlo input parameters.

breakdown of the contents that fall within these six parameters are also listed in Table 10.2. These points are not covered in detail, suffice to say that to achieve an accurate prediction plan, it is recommended to take these parameters into account.

10.4.2 Monte Carlo Iterations

A Monte Carlo iteration can be thought of as a computational procedure in which a cycle of operations or instructions is repeated a number of times to approximate and achieve the desired results. This consists of randomly distributing active UEs within a specified area, based on the active UE density map. The active UEs are then considered in both a sequential and random fashion within each iteration. The set of operations listed in the Monte Carlo algorithm flowchart in Figure 10.2 are performed for every UE, and are jointly performed for each UE currently present within the simulation. A passive scan terminal functionality should exist within any quality planning tool, which basically consists of offering the planner the ability to provide a full set of results after only a few snapshots have been performed. This is the basis of a 'snapshot calculation'. This functionality is advantageous and provides a fast solution for detecting network design problems caused by hard or soft blocking. Hard blocking occurs due to a lack of hardware, and soft blocking, when a user is blocked due to the cell running out of capacity. When the most obvious problems, such as the pilot power settings, the channel resource allocations, and noise rise requirements are solved, it is beneficial to run a large number of snapshots (pilot power is the power required for the pilot channel, which is an unmodulated code channel, cluttered with the cell-specific scrambling code, and can be thought of as a 'beacon'-type channel, similar to the old GSM broadcast control channel). This will ensure that statistical confidence can be achieved from the results presented by the passive scan functionality within the tool, as mentioned earlier. The exact number of snapshots required will be dependent on the proposed coverage area being simulated, and how many UEs, cells, services, and carriers are to be included. A point is reached where the overall simulation results minimally change from one snapshot to the next, and this is where statistical 'convergence' is achieved. A passive scan UE does not simulate the UE actually in a transmit mode, or nor does it simulate establishing a connection to the network, so the randomly scattered UEs that already

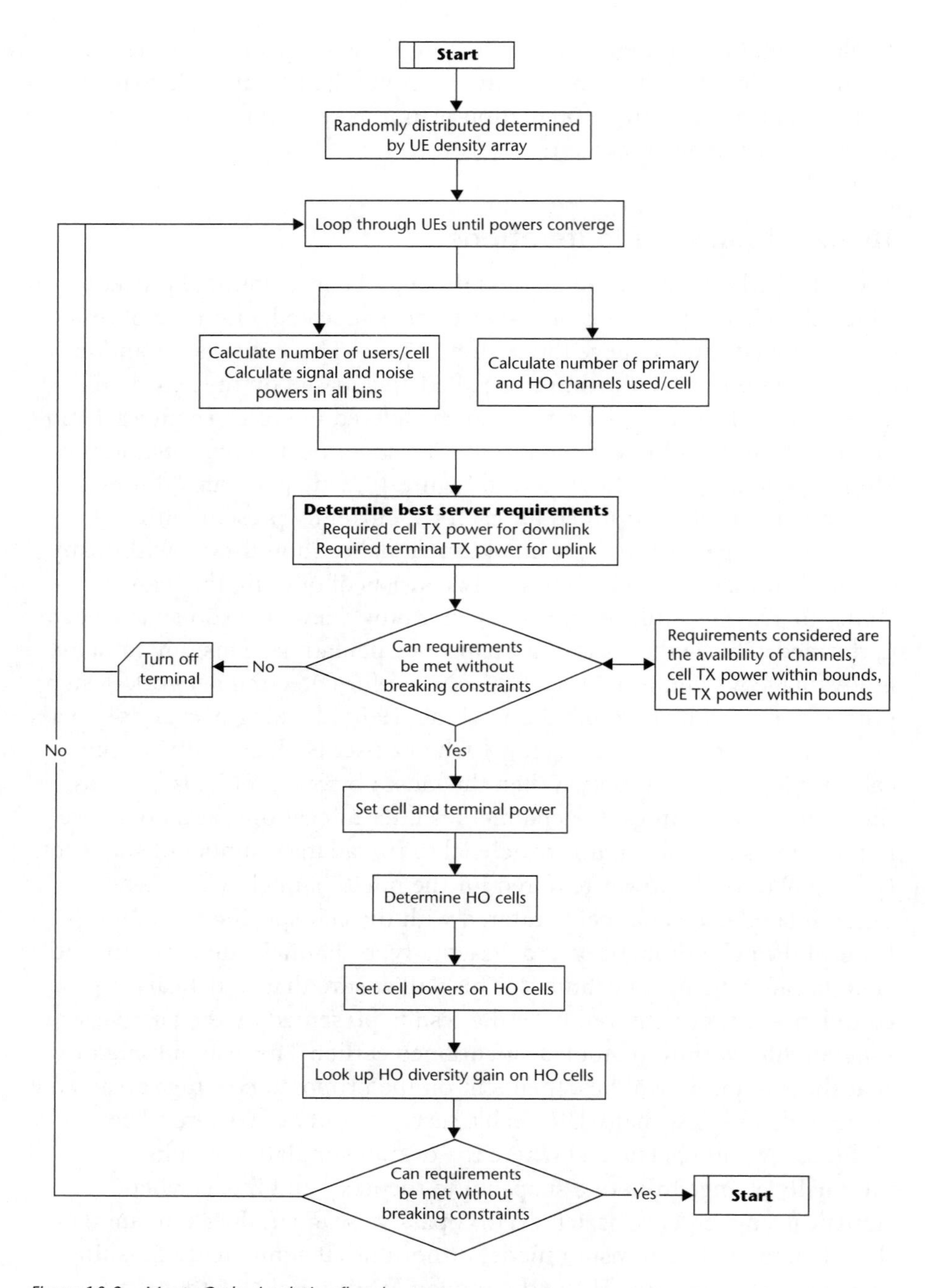

Figure 10.2 Monte Carlo simulation flowchart.

exist within the snapshot are not affected in any way. In addition, performing passive scans should not influence the end results, as the values in these cell reports are achieved merely by the randomly distributed UEs in the simulation.

10.4.3 Monte Carlo Results Analysis

The functionality for the analysis of both the coverage and interference on the uplink and the downlink should be available within the planning tool. This can be achieved through the creation of coverage and interference arrays, based upon the results of the Monte Carlo simulations. The results from these arrays should be able to provide the following results:

- Handover probability
- Coverage probability
- Reverse coverage probability
- Downlink total interference power, with respect to the UE effective radiated power (ERP)

10.4.4 Propagation Models

Within the planning tool itself there exists a choice of RF propagation models, the selection of which will depend upon the specific area and clutter present within the proposed planning area required.

The type of propagation model is heavily dependent on the actual clutter model used. Clutter can be defined as the morphology associated with urban and dense urban environments, namely buildings, etc. within the area where the RF will propagate. It is essential to account for building clutter when estimating any radio waves propagation characteristics. Certain sources of clutter may not utilize the same classification, for example urban, suburban, village, open, industrial, water, sea, forest, as opposed to residential, block buildings, parks, forest-dense, forest-light, or even semi-natural vegetation.

Macrocell models are well known and are normally based on statistical techniques. In practice, few planning tools use a single model, and most are a hybrid of the methods used in the various propagation models listed below, with a brief outline of their characteristics indicated in the following list:

1. Empirical model (statistical)
   - GSM standard – well established
   - Lower computational overhead
   - Less accuracy in small urban cells
   - Subjective clutter classification
2. Deterministic model (often known as the ray tracing model)
   - Not many proven models available as yet
   - Higher-computational overhead
   - More appropriate for small urban sites
   - Requires more extensive map data
   - Higher resolution
   - Higher accuracy
3. Street canyon model
   - Requires more extensive map data
   - More appropriate for microcells

10.4.4.1 Empirical Model

Empirical models analyse the vertical plane between the transmitter and receiver, and the parameters of the model are then gained from the properties of the buildings in this vertical plane. Two of the most common and popular empirical models used today are the Okamura–Hata model and the Walfisch–Ikegami model; these are really classed as the industry standard models. Unfortunately, they are not so accurate for smaller cells under 1 km, and the upper frequency limit for these tools is 2 GHz. However, these model types should not require any calibration and are more suited for urban macro- and microcells. Table 10.3 lists some of the popular microcell models.

10.4.4.2 Deterministic Models (Ray Tracing)

A deterministic model utilizes a method known as ray tracing, which can be performed in both the vertical and horizontal planes. Ray tracing can be

Prediction model	Method used	Parameters	2D	3D	Over roof top	Required data	Results
University of Lund	Dual slope empirical	–	✓	X	X	2D building layout	Pathloss
University of Karlsruche	Ray launching	Reflected and penetrated rays	✓	X	X	2D building layout	Pathloss + CIR
CNET	Ray tracing	Nine reflections and ground and street diffraction	✓	X	X	2D building layout	Pathloss
Swiss Telecom	Ray tracing	Reflection and diffraction	✓	X	X	2D building layout	Pathloss + CIR
Ericsson	Mathematical	–	✓	X	Walfisch–Ikegami method	2D building layout	Pathloss
Telecom	Street canyon	–	✓	X	Walfisch–Bertoni method	2D building layout	Pathloss
Walfisch–Ikegami	Semi-empirical	–	✓	X	Walfisch–Ikegami method	2D building classes	Pathloss
University of Valencia	Ray tracing	Scattering and diffraction	✓	X	Walfisch–Bertoni method	2D building layout + building height	Pathloss
CNET	Ray launching	Reflection and diffraction	X	✓	X	2D building layout	Pathloss + CIR
ASCOM ETH	Ray tracing	Reflection from wall and street	X	✓	X	2D building layout + building height	Pathloss + CIR
University of Stuttgard	Ray launching	Six reflections and two diffractions	X	✓	X	2D building layout + building height	Pathloss + CIR
University of Karlsruche	Ray tracing	The most dominant rays are considered	X	✓	X	2D building layout + building height	Pathloss + CIR

Table 10.3 Popular microcell models.

thought of as the RF diffractions and reflections that occur in the terrain and buildings, where the RF signals are propagating. Therefore the model's algorithms are able to take into account individual building footprints and heights, including specific terrain profiles. This is possible as the algorithms used are based on the uniform theory of diffraction and ray tracing. Diffraction describes the amount that a signal propagates over and around an obstruction. In general, the larger the obstruction the greater the diffraction. This tool is able to trace rays from each site through an accurate three-dimensional (3D) representation of the required area. Diffractions, absorptions, and reflections can also be incorporated within the model, thus providing even more accuracy. Some of the few deterministic models currently available are not so well accepted.

10.4.4.3 Street Canyon Model

With regard to street canyons, a deterministic model as described above will also be utilized, as these provide particularly good accuracy for areas of around 200 m or less from the base station. By employing ray tracing, such deterministic modelling is able to cope with small-scale fading or delay spread within the street canyon environment.

10.5 Mapping Data Requirements

Heights and clutter are the main requirements for all mapping data and the resolution is dependent on the following:

- Area type
- Propagation model type
- Budget

With regard to the map data, for urban area types, higher-resolution data is required (1–50 m). For rural areas, lower-resolution data is required (50–200 m). Microcell models require more detailed data than macrocell models. In general, macrocell models typically require around 20–200 m, whereas microcell models will need around 1–20 m. The reason behind this is that the better the quality and resolution of the purchased map data, the more accurate planning results can be achieved.

10.6 Carrier Wave Measurements

With regard to the previously mentioned propagation models, it will be necessary to perform some test measurements, thus enabling each type of model to be calibrated. Once these measurements are performed and inputted into the tool, this will ensure accuracy and confidence in the propagation predictions are achieved.

An average network requires three different propagation models as described previously, which will be used for planning in urban, suburban, and rural areas. It will then be necessary to perform measurements for each model, from around 10 to 15 sites to enable accurate calibrations to be performed. The test sites chosen for propagation modelling should be representative of a typical cross-section of cellular sites and should be free of any obstructions as this will result in inaccurate measurements. Global positioning satellite (GPS) fixes can be used to provide greater accuracy along with signal strength measurements performed over various points along the required routes. The test routes need to be planned carefully and should avoid elevated sections of roads, cuttings, tunnels, and bridges. To ensure the validity of the model, sufficient measurements must be made within each clutter type. Generally, the distances driven for each site would be around 80 km per urban test site, and around 160 km per rural test site.

In summary, static calculations, and static and dynamic simulations form the basis of the current planning methodologies available. Coupled with the introduction of Monte Carlo simulations, and the different propagation models, this should enable the planner to select the required model for the specific environment to be planned. Once this has been appreciated, the planner is in a position to digest the following chapter, which covers the nominal planning requirements and the actual site selection.

Nominal Planning and Site Selection

Nominal planning can be considered as the basis and initial starting point of site selection and should be regarded as one of the most important phases of the radio network deployment process. With vigilant and skilled planning, the required coverage, capacity, and quality of service (QoS), abbreviated as CCQ, can be achieved from the beginning of the radio network's life cycle, and simultaneously enable an adaptable network topology for future growth.

11.1 Nominal Planning

A nominal plan is initially classed as a hypothetical network or an imaginary network whereby prospective cells are theoretically located within the desired coverage area. This is considered as the starting point for the network roll-out process, and will then evolve and develop into the final network design. As physical sites are identified and acquired, the nominal plan is amended. A simplified flowchart illustrating the key processes can be seen in Figure 11.1, in which the first four elements shown in the diagram depict the nominal planning process.

11.1.1 Initial Network Dimensioning

Network dimensioning needs to be performed in order to ensure sufficient capacity is available to carry the required traffic. With initial dimensioning

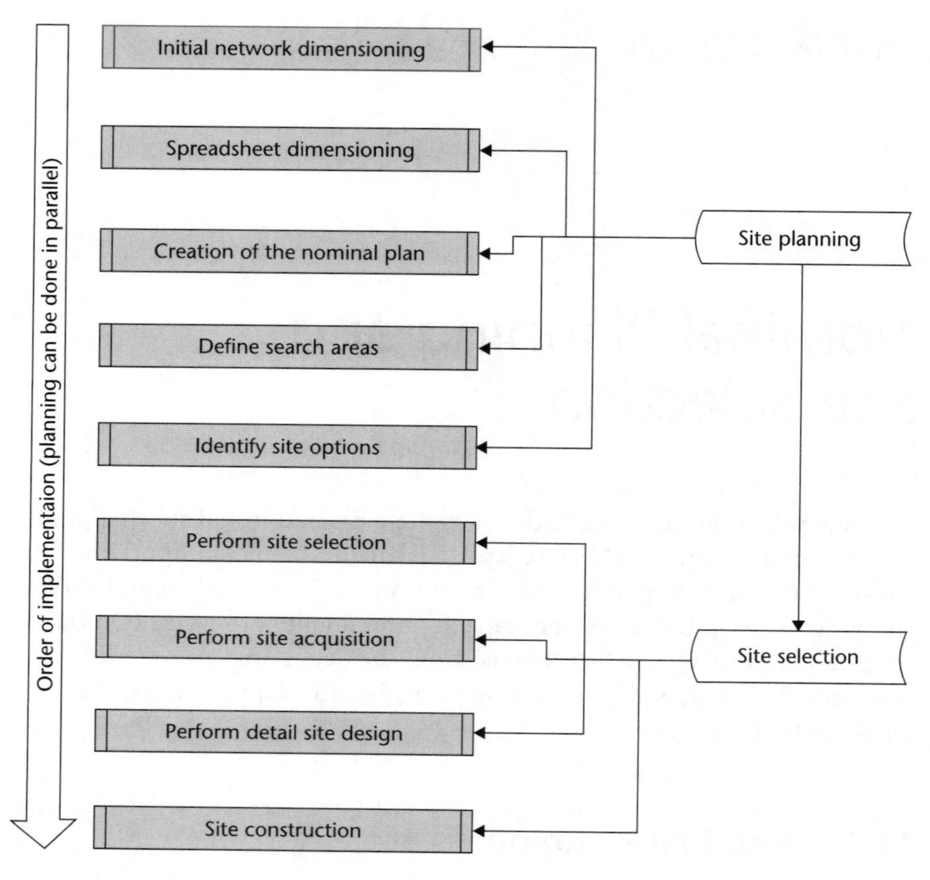

Figure 11.1 Nominal planning.

a predicted number of sites and network elements will form the initial plan towards securing an accurate image of the required network. In addition, an increase in the performance of the network should be taken into account, thus highlighting the need for a correct determination of the design parameters as well as the architecture of the system and the traffic requirements.

From the initial dimensioning process it should be possible to identify the approximate number of sites required for the desired coverage area. The identification of the approximate site radii required for urban, suburban, and rural areas can also be achieved. This will include both voice and data services as shown in Table 11.1. This should be used as a major input to the nominal plan as ultimately it will enable more accurate results to be achieved. A more detailed table indicating typical ranges for universal mobile telephony system (UMTS) sites can be seen in Table 6.12 (in Chapter 6).

Table 11.1 Initial dimensioning.

	Urban (km)	Suburban (km)	Rural (km)
Voice	1.8	3.1	4.4
64 kbps	1.6	2.7	3.5
384 kbps	1.1	2.4	3.2

■ Specific service; ■ Maximum range required to support all services; ■ Service not supported in this environment.

The examples listed (in Table 6.12) can be used as a guideline for both indoor and outdoor coverage probabilities within certain ranges at standard perceived data rates. Sample ranges are also given for both real-time (RT) and non-real-time (NRT) traffic. RT traffic is unable to withstand any delays and hence requires a high level of QoS, as previously explained in Section 7.3; examples of such types of traffic could be video-to-video calls and live multimedia connections, etc. NRT traffic is able to withstand delays and could consist of such traffic types as web downloads and e-mails, etc.

11.1.2 Spreadsheet Dimensioning

A spreadsheet dimensioning method can be used to assist in the nominal planning stage and will be able to model the capacity and coverage requirements (see Figure 11.2). A coverage-driven spreadsheet can be constructed along with a capacity-driven spreadsheet and a combined approach can then be modelled. This method can be used as a complex coverage versus capacity trade off tool. With regard to coverage, the link budgets can be used to calculate the maximum pathloss. Once this has been performed the results can then be converted into cell radii that can depict different types of environments, such as urban, or suburban areas; finally estimates can be performed for the average site coverage area required for each type of environment.

It should be noted that the number of sites per environment is equal to the environment divided by the average site coverage area. The number of sites per environment for each zone or coverage area can be added together, and from this data assumptions on cell loading and services can be made within the link budget process.

Spreadsheet dimensioning tools for 3G should consist of a combined iterative approach:

- Cell range is calculated from the link budget which contains an interference margin calculated from the cell loading

- The cell coverage area can then be calculated from the range

- The area and subscriber density can then be used to calculate the volume of traffic captured by a cell

- Using the captured traffic, the loading of the cell can then be calculated, and the link budget can then be re-calculated from the new loading

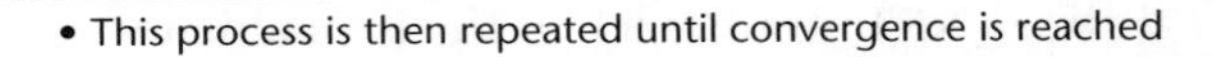

- This process is then repeated until convergence is reached

Figure 11.2 Spreadsheet dimensioning.

With regard to capacity, if an estimate of the traffic profile per subscriber is made, then the offered traffic per square kilometre per environment can be calculated. Given the capacity of a cell, an estimate can be made for the average site capacity per environment. Hence, the number of sites per environment type is equal to the environment area divided by the average site capacity area.

The number of sites required per environment for each zone can then be added together.

$$\begin{array}{c}\text{Number of sites}\\ \text{(per environment)}\end{array} = \frac{\text{Environment area}}{\text{Average site coverage area}} \tag{11.1}$$

From performing an analysis such as that described in the previous paragraph the results can be combined, and a comparison for the number of sites required per zone derived from the coverage and capacity approach. From making this comparison it should be possible to assess if the zone is either capacity or coverage constrained, so the output of the dimensioning would be the largest number of sites per zone. This type of dimensioning is also likely to be used by vendors attempting to estimate the number of sites required to fulfill a bid proposal for a prospective network operator. One of the disadvantages with the spreadsheet approach, is that it is unable to provide the user with a fair visualization of how to accurately interpret the results. Problems are likely to occur with microcells, road coverage, urban overlapping, and transmission dimensioning. This type of dimensioning is 'acceptable' for global system for mobile communications (GSM). However, the coverage and capacity relationship in third-generation UMTS (3G) is much closer, therefore typically spreadsheet dimensioning tools for 3G will need to take a combined iterative approach. The steps are illustrated in Figure 11.2 and require an iterative process to be executed until convergence is reached. Convergence can be described as the point where the overall results of the previous iteration are either equal to or only indicate a marginal difference to the final iteration.

11.1.3 Creation of a Nominal Plan

The creation of a nominal plan involves the positioning of a hexagonal grid over the desired coverage area as shown in Figure 11.3. When the radius of

Figure 11.3 Creating a nominal plan.

each hexagon can be determined, then it will be possible to attain an idea of the predicted capacity of the planned network. Following this, it will then be possible to detect any 'hot-spots' that may require cell splits and any under-utilized cells that may not be required and could possibly be removed. Cell splits can be particularly advantageous and can be a replacement for the omni-directional cell which has an antenna radiating equally in all directions with several directional antennas on the same mast. This 'sectorizing' causes the previously homogeneous cell to be divided into three or six distinct areas (each with either 120° or 60° coverage around the site, respectively). Under-utilized cells consist of cells that are unlikely to carry a reasonable amount of traffic, therefore bringing into question whether an entire cell can be justified in such a low volume traffic area. This can usually be resolved by removing the under-utilized cell and attempting to cover the required location by boosting the output and upgrading the neighbouring cell sites to compensate. A survey of each nominal site would then be performed to identify possible site options. It is usual practice to pursue three possible site locations, known as options, explained in Section 11.1.5. Guidelines must be drawn up so that the site surveyor will be able to determine all options that will provide the required radio coverage. This will involve analysing the terrain and the height of structures and buildings, etc. Such an analysis would normally be provided in the form of a search area, which could be the actual radius from the nominal site, or possibly one or more polygons following existing map height contours. Ideally, an established operator would start with the existing GSM sites, evaluate the coverage and then add in the areas without coverage, often known as gap fillers. Once the existing GSM sites have been considered, the possibility of major site owners should be exploited, for example the railway and utility companies. All options must meet the required criteria as described in Chapter 11, and be both viable for construction and technically acceptable.

11.1.4 Defining Search Areas

Due to the requirements of the system it is likely that for UMTS (3G) networks a larger number of sites will be required than with previous cellular networks. This has the effect of increasing the constant pressures of identifying smaller search areas, and consequently a reduction in the availability of suitable buildings within that area. In particular, council planners and property owners are far more concerned with the aesthetic issues for future developments than with the previous GSM cellular sites.

On the positive side, technology is also evolving, thus enabling the use of smaller equipment and antennas. However, operators have to think carefully when deciding their approach to locating antenna masts. Most planning departments and local communities are concerned about the appearance of masts, particularly in residential areas. Equipment and mounting installations can now be concealed more effectively than before so that they blend in with the environment. Furthermore, specific camouflage and stealth sites options are being designed, for example, equipment located onto closed circuit television (CCTV) columns or lampposts along with false panelling and new materials that can be moulded into the required shape and painted, thus increasing the camouflage potential. Due to the availability of these modifications, local authorities have been, and are more likely to be more positive, hence these points should be considered when attempting to define a suitable search area.

11.1.5 Site Options

Evaluation of both the radio coverage and the transmission back haul enabling the traffic to be relayed through the network should be performed, along with the selection of both a preferred and a back-up site choice, known as a site option. After these site options have been agreed, the site design elements must be considered; following this the site acquisition process can be started. It is highly advantageous in both time and cost if the acquisition agent and the site designer work in parallel as this can rule out any unfeasible sites with regard to structural suitability. This then means that when the options are presented they consist of three practical alternatives.

The site acquisition agent should prepare a report for each site listing the possibilities available. This report will include the accurate grid reference, the height of structures or available antenna windows, site photographs and 360° panoramic photos from the site. Initally three options should be pursued. The additional options cost more at the outset, but when complications with the chosen option occur, they will prove cost-effective, preventing new searches, structural reviews, and expensive delays.

Following the selection of the three options the site acquisition department and the site designer can begin the site negotiations, planning and/or permit drawings, leases, and the planning permission process. Taking into account the new option of stealth and camouflage site development, photomontage presentations are also developed at this stage.

11.2 Perform Site Selection

When performing site selection various factors must be considered such as acquisition, logistical issues, power availability, structural design and integrity, accessibility, and aesthetics all the way to determining whether potential sources of high level radio frequency (RF) signal interference exist. Having three options available, as discussed at the end of the last section, helps to minimize these problems. As with the CCQ model, it is a balance of all these items against the overall cost of the installation that aids in the selection of the optimum cell site.

It is at this stage that we perform the last four stages of the nominal planning process as shown in Figure 11.1 and it is these requirements that are now reviewed so that the most favourable site from the given cell site options is selected. The following criteria in this section will assist in resolving a definitive cell site: transmission, access, power, and planning requirements as described in further detail below.

11.2.1 Radio Requirements

For each cell option and the surrounding cell sites a RF static calculation should be performed, thus ensuring that each potential cell site will provide sufficient coverage for the target areas required. It should be noted that in this instance a static calculation refers to a RF static calculation as described previously in Chapter 10. In addition, it is important for the radio emissions from each site to be 'contained' within the required target coverage area. Finally, any obstructions that might create problems with regard to the RF emissions must also be evaluated. The remaining potential cell sites should be considered on their projected coverage, their interference potential and the distance from any potential hot-spots. To avoid major overlap or coverage holes, the location of adjacent cells site options should be taken into account.

11.2.2 Transmission Requirements

From previous experience it has become apparent that transmission delays often impede the network roll out process, thus taking longer than expected to resolve. These are usually a major issue and the choice of site often impacts on the transmission plan.

It is beneficial to perform a preliminary evaluation with regard to ensuring a line-of-site (LOS) availability to the required microwave transmission nodes. If due to site location this is not possible then duct and riser availability for leased line and fibre routes within the chosen building must be ascertained, offering an alternative transmission route.

11.2.3 Access Requirements

Access requirements are in two stages: initially the site needs to be accessed for construction purposes, following which the site needs to be accessed for ongoing operational maintenance and future upgrades.

Access for construction is normally straightforward for a greenfield site with the additional costs usually being the machinery used to get the equipment to site. This could, for example, include specialized cranes or temporary access tracks, in order that difficult terrain (steep or muddy tracks), or access in confined areas (dense forests with no access tracks) can be negotiated. Location and access must also be considered in certain locations during changing seasons, for example sites in mountainous areas can often become inaccessible both to the site and into the cabin due to a snow barrier. For similar reasons flood zones are to be investigated, especially when considering pylon sites.

Rooftop sites can have their own access problems during construction, but problems mainly occur in the long term. With regard to rooftop construction access, in extremely difficult locations where cranes cannot reach or get permission to operate, helicopters are used to deliver the equipment. However, this only usually occurs when all other options have been exhausted or if the site is a prime site due to the excessive cost of installation and so should be considered as a last resort.

Operation access is considerably more problematic on rooftop locations than greenfield. In most cases, especially prime locations, access needs to be 24 hours, 7 days a week and therefore access to offices and apartment blocks, government buildings and even churches needs to be simple and safe. This latter aspect can now add a substantial cost to the overall implementation if additional safety ladders and walkways are to be installed.

This has been just a short introduction into the problems of site access, however, its purpose is to ensure that access is not overlooked when selecting and planning a site due to the implications for safety, cost, and maintenance that can occur both in the short and the long term.

11.2.4 Power and Planning Requirements

Without power the site will not operate and while during the initial rush to implement sites it is possible to use temporary generators, in order to achieve coverage, this is not advisable without planning the power route. In certain remote locations the delivery of the power supply can be the biggest part of the site development costs. Furthermore, temporary power suffers from reliability, servicing, and safety issues.

A suitable power supply is also dependent on the quantity of the delivery and on the fact that excessive demands on a standard supply can result in costly upgrades to the supply, or excessive charges due to a third party supplier such as the landlord.

When planning a site it is always an advantage to obtain a general development order (GDO) or equivalent as sites can be built with minimal restrictions as long as certain criteria are met. The criteria can include height and location restrictions, but can differ from country to country. The alternative is to obtain full planning permission which is not only a time-consuming exercise that needs to be planned and managed in advance, in order to meet implementation targets, but obtaining permits can also be a costly exercise requiring the submission of documentation which includes survey details and a full set of planning drawings. Furthermore, there are the potential political and aesthetic issues that may need to be considered, for example sites located near to hospitals and schools, or in, or next to, conservation areas.

The above has highlighted the importance of site selection in terms of acquisition, detailed site design and some of the problems of actually constructing the site. However, these aspects are not to be considered alone and the following section reviews the problems associated with sites inadequately designed in terms of the radio plan.

11.3 Effective and Ineffective 3G Universal Mobile Telephony System Sites

Efficient radio engineering practice becomes more critical with UMTS, as control of the RF emissions is vital for a successful network. The main objective is to choose a site location that provides the required coverage, but also ensures the RF emissions are kept to a minimum. Ideally the site should be located as close as possible to the predicted traffic, so that less output power will be required by both the user equipment and the base station. Minimizing the output powers will in turn ensure that the noise affecting other users on both the serving cell and the other nearby cells is also minimized. It should be noted that only very minimal coverage can be achieved in the case of a site being deficiently located with respect to the existing terrain/clutter, along with the antenna heights being too low.

In contrast once the antenna height is raised too high then the RF will propagate further, spilling over into other regions populated by other adjacent cell sites, thus causing interference within these adjacent cells. This in turn will reduce both the range and the capacity of these adjacent cells.

The type of environment in which the site is to be located, for example dense, urban or rural, will determine the actual antenna heights. One of the most important factors to be kept in mind with 3G planning is how well the propagating RF emissions can be controlled.

11.4 Suggested Antenna Heights

Antenna heights will prove to be a key issue with regard to achieving the required UMTS coverage. As most operators already have a working GSM cellular network, UMTS will be implemented together with these existing GSM networks. It is currently predicted that an existing operator will only be able to utilize around 30–50 per cent of the existing GSM sites to co-locate UMTS equipment. A high volume of existing GSM sites will not be suitable due to the height, terrain, power availability, load, aesthetics, radio frequency interference (RFI), and in addition lack of actual physical space available on the site for the equipment, especially with certain multi-operator sites. Co-locating with other operators is nevertheless strongly recommended. There are certain advantages to be gained from this, for example co-operative site owners, easy planning permission, power availability, and transmission routes. However, there are some disadvantages, such as that the shared sites have not been designed to fit into the nominal plan and probably will not provide optimal coverage. Wind loading and tower strength issues may also arise with certain sites, there may be problems whereby numerous microwave point-to-point LOS links have to be located on a single mast, and finally there are possible interference problems. Some guidelines for antenna heights with respect to different types of environments are given in the following subsections with a view to achieving optimal coverage.

11.4.1 Dense Urban Sites

With regard to actual antenna heights, for dense urban areas the antennas should be located below the average clutter height for the area. The optimal height will depend on the actual clutter height of the specific environment to be covered, therefore no actual example height figures can be stated and only guidelines given. However, this will have a beneficial effect of using the clutter to protect the site from far field noise. Far field noise can be thought of as an unwanted radiating RF signal that is occurring further away than the wanted RF signal's wavelength. The area past this distance is known as the far field and can cause unwanted RFI. This in turn will cause a reduction in RF emissions, which have a detrimental effect in causing noise to all adjacent sites. In addition, it is considered advantageous to locate antennas on the side of buildings dependent on their heights, as this will suppress unwanted antenna side and rear lobes. As all antennas possess directional qualities, they do not radiate equally in all directions, hence RF will emanate from both sides of the main wanted beam and in addition also to the rear of the main beam. These are known as the side and rear lobes, and in a well-designed antenna they should typically be 10–15 dB below the main beam. It may be that a dense urban site of this nature may not have a relatively good unobstructed 'RF view' to most of its coverage area, so the planner should be aware that any blockages occurring due to nearby buildings will cause an unwanted reduction in coverage.

11.4.2 Light Urban and Dense Suburban Sites

In light urban areas antennas should be located just above the average clutter height for the area, hence a rough guideline would be up to 5 m above the average clutter height. This will act in a similar manner as described above whereby the site will be masked from far field noise and the RF emissions will be kept to a minimum, thus reducing interference to neighbouring sites. Generally a reasonably unobstructed 'RF view' should be available to most of the cells required coverage area, but not much further than the cell periphery.

11.4.3 Light Suburban Sites

In areas where lower traffic density is expected such as light suburban sites, the antennas should be located around 5–10 m above the average clutter height, which will vary as each specific environment will be different (e.g. different tree heights, etc.) The same criteria apply for the light suburban sites whereby again, a reasonably unobstructed 'RF view' should be available to most of its coverage area, but not much further than the cell periphery.

11.4.4 Rural and Highway Sites

For rural and highway sites the antennas should be located around 10–20 m above the average clutter height, whereby a relatively unobstructed 'RF view' from the site exists in all directions. It is likely that the RF power radiating from these particular sites will be of a higher level than sites located in the previously mentioned environments; therefore it will prove more difficult to effectively control the RF emissions. In this case it is possible that unwanted RF may spill over and reach a town or city. Therefore, to control these emissions, a method of both electrical and mechanical antenna tilting will need to be implemented. This has been covered in more detail in Chapter 12.

11.5 Traffic Calculation

To evaluate the cell capacity, firstly it must be assumed that the packet data (NRT traffic), can be scheduled to fill the remaining RT capacity without exceeding the pre-determined peak traffic load. When all the spare RT capacity has been utilized, the remaining capacity must be then converted to 12.2 kbps voice equivalent circuits. Once this has been performed, an estimate can be made as to whether the capacity of the cell has been exceeded.

As shown in the calculation illustrated in Figure 11.4, for this example it is assumed that a cell with a capacity of sixty 12.2 kbps voice circuits captures:

- 1.59 Erlangs of voice traffic.
- 0.69/384 kbps data users.

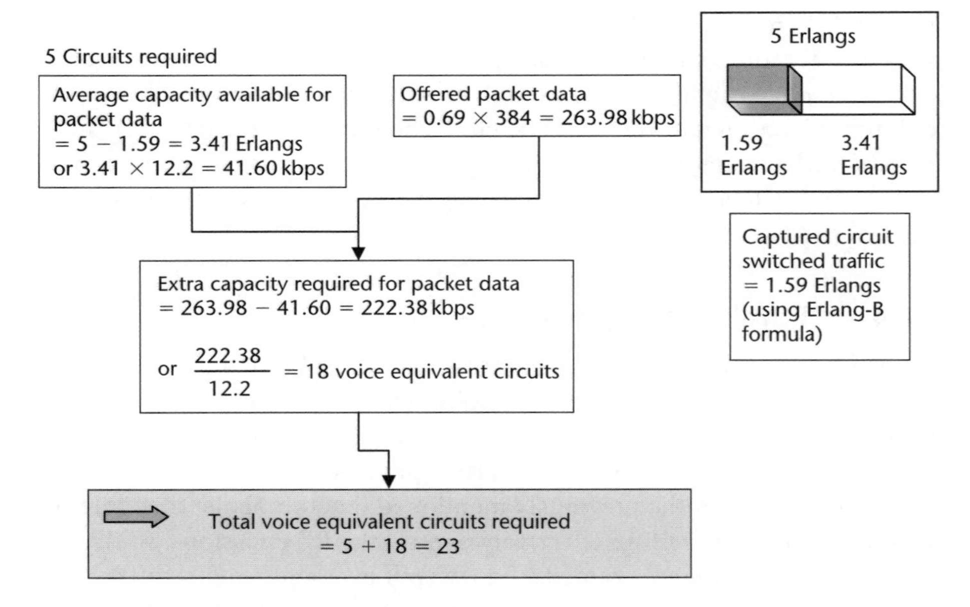

Figure 11.4 Evaluating traffic requirements.

Traffic requirements must be evaluated within a 3G system, as it is important to ensure the load can be effectively managed.

11.5.1 Traffic Evaluation

In the traffic evaluation example shown in Table 11.2, five cells are predicted as being overloaded, and one cell is close to being overloaded.

Sites	Voice (Erlangs offered)	Voice (circuits required)	Voice equivalent circuits available for packet data	Data rate available for packet data (kbps)	Number of 384 kbps packet data users	Data rate required for packet data (kbps)	Extra capacity required for packet data (kbps)	Extra voice equivalent circuits required for packet data	Total voice equivalent circuits required	Overload status
A1-1	1.27875	5	3.73	45.41	0.62	237.5	191.85	16	21	OK
A1-2	18.989	27	8.02	97.72	9.13	3507.56	3409.84	280	307	Overload
A1-3	2.6418	7	4.36	53.18	1.24	474.92	421.74	35	42	OK
B1-1	18.0142	26	7.90	96.33	9.05	3476.01	3379.67	278	304	Overload
B1-2	0.099755	2	1.90	23.18	0.05	19.15	0.00	0	2	OK
B1-3	1.71587	6	4.28	52.27	0.77	295.33	243.06	20	26	OK
C1-1	2.13376	6	3.87	47.17	0.73	281.12	233.95	20	26	OK
C1-2	1.58312	5	3.42	41.69	0.69	263.98	222.29	19	24	OK
C1-3	105.62	118	12.94	157.84	52.44	20137.08	19979.23	1638	1756	Overload
D1-1	11.8475	19	7.15	87.26	5.72	2195.91	2108.65	173	192	Overload
D2-2	2.43671	7	4.56	55.67	0.96	370.13	314.46	26	33	OK
D2-3	3.5485	8	4.45	54.31	1.75	670.97	616.66	51	59	Overload

Table 11.2 Traffic evaluation.

In this case all six cells will need to be split to avoid congestion. Any omni sites that are congested should be sectored, and overloaded sectored sites need to be off-loaded onto new cells. After the nominal plan has been modified accordingly, care must be exercised to ensure the network is correctly dimensioned, hence the capacity analysis must therefore be repeated.

In summary, most of the issues covered in this chapter can be considered as the initial starting point with regard to cell planning, taking into account how to perform site selection, consideration of radio and transmission requirements, and the many and various peripheral factors that need to be taken into consideration when optimizing either a single or cluster of sites. Further optimization can be achieved through the micro-aspects of site design covered in the following chapter.

CHAPTER TWELVE

Optimization through Detailed Site and Antenna Configuration

12.1 Introduction

This chapter covers fundamental issues with respect to site and antenna configurations. This in turn will ensure that optimal coverage, capacity, and quality of service (QoS), abbreviated as CCQ, can be achieved, along with ensuring cost savings can be made by employing the correct types of site configurations within the different types of environments. Optimization issues are also covered, encompassing uplink and downlink coverage gains, isolation requirements, and co-siting and tilting of the antennas.

12.1.1 Basics of Site and Antenna Configuration

All antennas are defined mainly by their radiation characteristics. Radiation patterns can be thought of as two-dimensional images, which illustrate how much energy the antenna is able to radiate and in which direction. A distinct relationship applies between the directivity of the radio frequency (RF) pattern and how the antenna focuses the energy.

To simply quantify the shape of the radiation pattern, which can be thought of as a graphic representation of where and how much energy is radiated, the half power beam width, side and rear lobes, and main beam tilt need to be defined, and are explained in further detail in this chapter. The antenna gain can be thought of as the actual efficiency of the antenna and to some extent by the return loss or the variable standing wave ratio (VSWR). In addition, the polarization describes how the radiated energy propagates through space and is also explained in further detail in this chapter.

As the antenna's propagation pattern must be accurately predicted, in turn the location, obstructions, antenna type and configuration, cable runs, and interference issues are some of the parameters that need to be taken into consideration.

12.1.2 Basics of Optimization and Configuration

As with global system for mobile communications (GSM) networks, optimization must be taken into account. In essence this can be described as a measure of the performance of the radio network. This performance can be gauged by radio engineers performing field trial measurements. The results of these measurements will enable the operator to determine the overall performance of the radio network as seen by the subscribers. Any problems caused by, for example, interference, low signal levels, and coverage gaps can be identified and optimization can commence.

12.2 3G Universal Mobile Telephony System Antenna Arrangements

As with previous GSM systems there are various antennas that can be utilized, and within universal mobile telephony system (UMTS) further varieties of antennas have evolved, and are covered in this section. Equipment manufacturers support the same basic antenna configurations: omni's, three- and six-sector sites. However, as expected, each equipment manufacturer supports his own variations on these configurations. Some solutions may require similar equipment configurations to a GSM base station and some may increase the number of antennas on a site. It is likely that eventually numerous configurations will exist due to the wide variety of UMTS antennas likely to become available. Initially three-sector sites will be employed, however a good example would be for a UMTS site to consist of six-sectors, whereby each sector would employ two antennas thus providing greater coverage by being able to use narrow beam antenna patterns, and hence in turn also increasing capacity. Polarization diversity antennas will be able to reduce this requirement to one antenna per sector. With polarization diversity, different directions may experience different fading. This applies especially to indoor environments, as polarization directions have been shown to be nearly uncorrelated, thus providing diversity, and in turn can be considered advantageous from a coverage point of view. The advantage of polarization diversity over antenna diversity is that polarization diversity does not require separation between antennas, and thus it can be applied to more compact equipment. However, there can be a power loss, for example, 3 dB in the downlink signal of the polarization diversity, as the transmitter power is being split into two polarizations. In addition, the specific performance of the polarized diversity antenna relative to space diversity is a function of the amount of reflection and/or scattering in the local environment. Field measurements suggest that the use of polarized diversity antennas in certain morphologies may result in a loss of performance up to 2 dB.

12.2.1 Transmit Diversity

It is important to consider transmit diversity as this method involves transmitting the same signal through two antennas. It can be best thought of as if several transmitters send simultaneously with full power an additive superposition of both signals can be seen at the receiving user equipment (UE). This performance gain can be attributed to the random superposition of the data signals at the receive antenna. The received signal power and the signal-to-noise ratio (SNR) are the summation of the receive powers of the single data signals with respect to the SNR of each single branch. In short, it can be stated that transmit diversity reduces signal-fading effects, which are detrimental to the system performance; therefore an increase in capacity can be achieved. This is particularly prevalent for the low speed, high data rate user.

As illustrated in Figure 12.1, if transmit diversity is utilized large gains or improvements can be achieved from both a coverage and capacity point of view. Transmit diversity consists of transmitting the downlink signal via

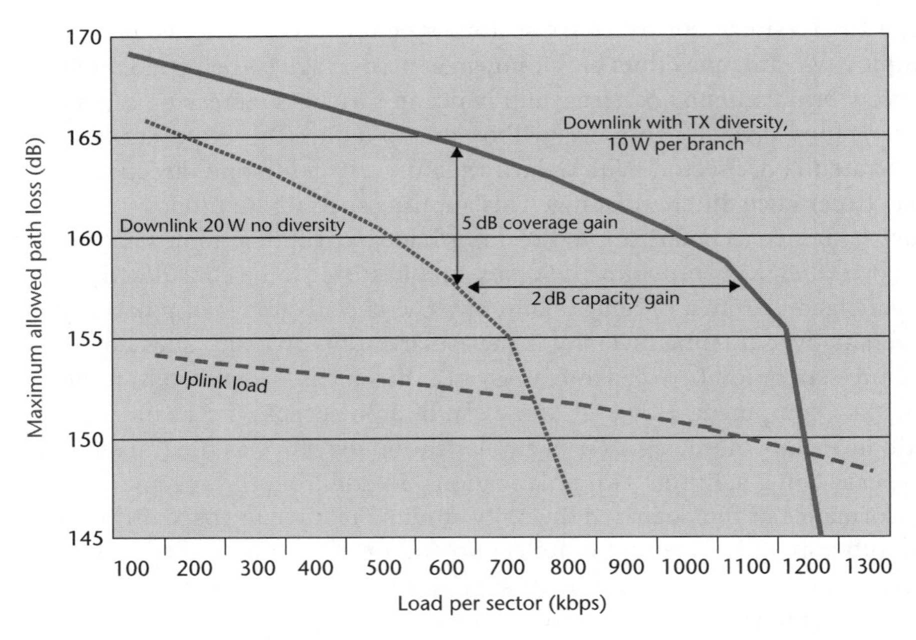

Figure 12.1 Downlink gains with TX diversity.

two base station antenna branches. If the base station already employs receive diversity, then it will be possible to duplex the downlink transmission to the receive antennas, and hence, no extra antennas for downlink diversity will be required. Downlink transmit diversity could use either polarization diversity or space diversity antennas. In addition, the advantages of utilizing transmit diversity consist of only using two antennas, which can only be considered to be beneficial from an aesthetic and cost point of view, and also with respect to CCQ, if transmit diversity is employed through transmit diversity antennas an increase in both coverage and capacity can be achieved.

12.2.2 Receive and Transmit Diversity

With the type of system illustrated in Figure 12.2 capacity improvements of up to 75 per cent can be realized, with roughly around 30 per cent of fewer sites required. This can be made possible because using multiple antennas makes significant improvements in both the signal performance and a reduction in interference levels. In addition between 2 and 8 dB improvements in signal strength can be achieved for both voice and data traffic, hence this in turn will require fewer sites to be implemented. This is four-branch diversity, however it reverses the trend to lower the profile of the site, for example the aesthetics, especially as each cell will require four separate antennas. Transmit diversity is therefore recommended, and this system can be implemented with two space separated dual polar antennas.

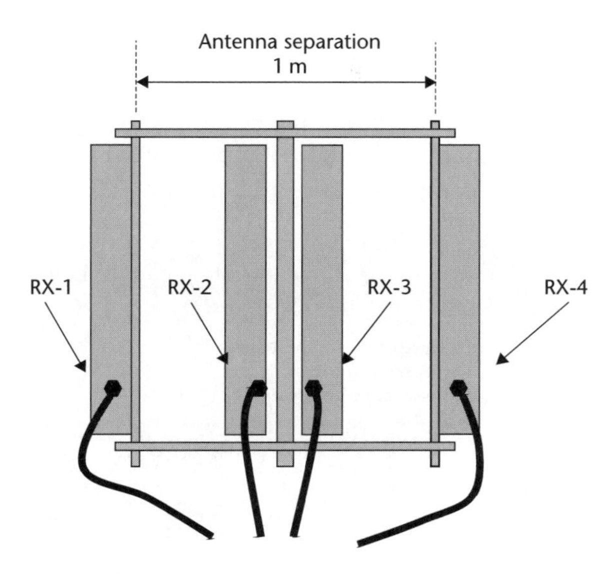

Figure 12.2 Four-branch diversity.

These so-called 'Smart' antennas consist of multi-beam or adaptive array antennas without handovers between beams. Smart antennas are also known as intelligent or adaptive antennas and possess the ability to 'track' the UE by having closely spaced pairs of antennas for beam steering. Multi-beam antennas use multiple fixed beams in a sector, while in an adaptive array the received signals by the multiple antennas are weighted and combined to maximize the SNR. This is made possible by utilizing antenna arrays that combine the signals to or from individual antenna elements. The angular position of the beam can be set by approximately combining the signals either to or from the separate individual antenna elements. So to summarize, coverage can be improved by the use of smart antennas.

12.2.3 Other Antenna Arrangements

The following subsection describes antenna systems that for the various reasons described would be beneficial for use within a UMTS network.

12.2.3.1 Dual System Antennas

There are many dual system antenna types available from a variety of manufacturers. The sample specification sheet illustrated in Figure 12.3 shows a dual band, dual polarized antenna with an 18 dBi gain. Note: the specifications are the same for both GSM and UMTS.

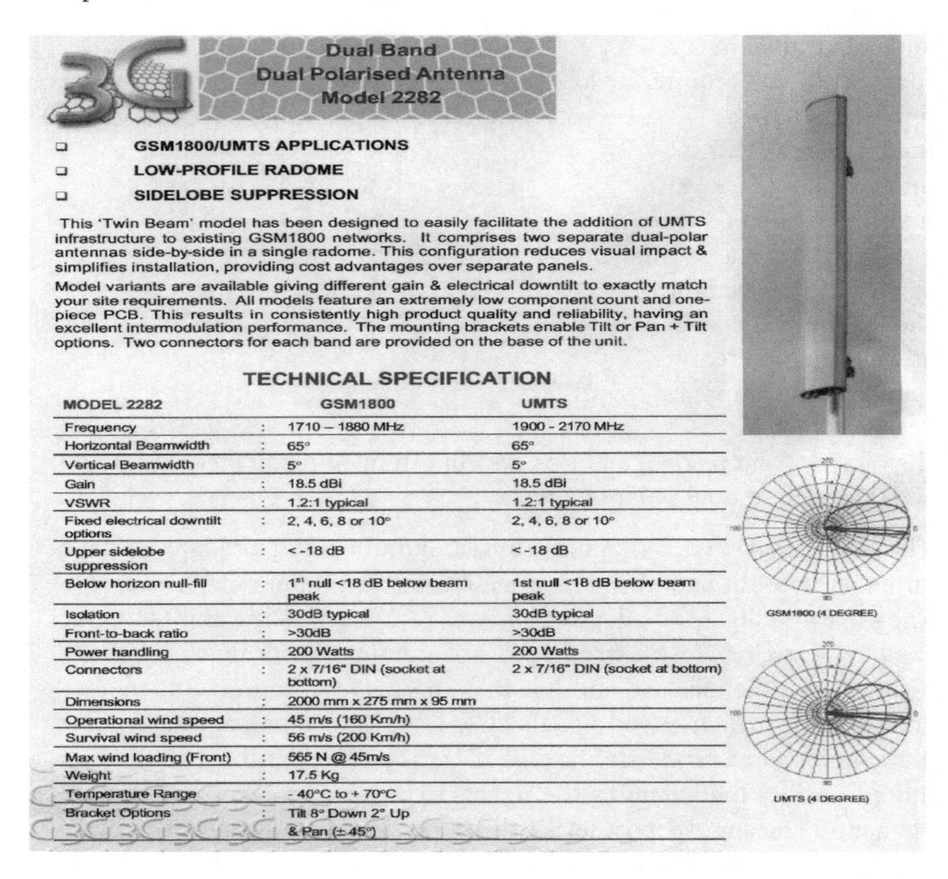

3G

Dual Band
Dual Polarised Antenna
Model 2282

- **GSM1800/UMTS APPLICATIONS**
- **LOW-PROFILE RADOME**
- **SIDELOBE SUPPRESSION**

This 'Twin Beam' model has been designed to easily facilitate the addition of UMTS infrastructure to existing GSM1800 networks. It comprises two separate dual-polar antennas side-by-side in a single radome. This configuration reduces visual impact & simplifies installation, providing cost advantages over separate panels.

Model variants are available giving different gain & electrical downtilt to exactly match your site requirements. All models feature an extremely low component count and one-piece PCB. This results in consistently high product quality and reliability, having an excellent intermodulation performance. The mounting brackets enable Tilt or Pan + Tilt options. Two connectors for each band are provided on the base of the unit.

TECHNICAL SPECIFICATION

MODEL 2282		GSM1800	UMTS
Frequency	:	1710 – 1880 MHz	1900 - 2170 MHz
Horizontal Beamwidth	:	65°	65°
Vertical Beamwidth	:	5°	5°
Gain	:	18.5 dBi	18.5 dBi
VSWR	:	1.2:1 typical	1.2:1 typical
Fixed electrical downtilt options	:	2, 4, 6, 8 or 10°	2, 4, 6, 8 or 10°
Upper sidelobe suppression	:	< -18 dB	< -18 dB
Below horizon null-fill	:	1st null <18 dB below beam peak	1st null <18 dB below beam peak
Isolation	:	30dB typical	30dB typical
Front-to-back ratio	:	>30dB	>30dB
Power handling	:	200 Watts	200 Watts
Connectors	:	2 x 7/16" DIN (socket at bottom)	2 x 7/16" DIN (socket at bottom)
Dimensions	:	2000 mm x 275 mm x 95 mm	
Operational wind speed	:	45 m/s (160 Km/h)	
Survival wind speed	:	56 m/s (200 Km/h)	
Max wind loading (Front)	:	565 N @ 45m/s	
Weight	:	17.5 Kg	
Temperature Range	:	- 40°C to + 70°C	
Bracket Options	:	Tilt 8° Down 2° Up & Pan (± 45°)	

GSM1800 (4 DEGREE)

UMTS (4 DEGREE)

Figure 12.3 Dual system antenna. Courtesy of Racal Antenna Systems.

It is likely that dual band antennas and feeder cable sharing between second and third generations (2G and 3G, respectively) will become a reality, since the timing and financial realities of implementing new 3G sites will out weigh the disadvantages of co-siting. The full costs required for a new site, coupled with the difficulty of acquiring them make it likely that the majority of operators will use their existing GSM sites in the first phase. Environmental issues will force many operators to consider deploying dual band antenna systems. In congested urban sites particularly, multiplexed feeder cable solutions may prove necessary for sites where the installation of new feeder cabling may not be viable.

With regard to CCQ issues, these three parameters will differ for each particular site and its type of environment, and will have to be assessed individually.

12.3 Ancillary Equipment

This section covers mast-head amplifiers (MHAs), otherwise known as tower-mounted amplifiers (TMAs), that can be used to provide CCQ enhancements in certain site configurations and environments.

12.3.1 Tower-Mounted Amplifiers

TMAs, sometimes referred to as MHAs can provide extra gain and are able to increase the receiving range of a base station. There are two stages by which the incoming signal is amplified, but if the signal is too weak the amplified results will essentially be useless due to the signal-to-noise ratio, which is often written S/N (or SNR as used), and can be best described as a measure of the signal strength relative to the background noise. This ratio is usually measured in decibels.

To improve the SNR the amplifier is placed in close proximity to the antenna, thus reducing unnecessary cable losses (see Figure 12.4). A typical low-noise amplifier (LNA) has a noise figure below 4 dB. The noise figure is the ratio of the actual output noise to that which would remain if the LNA itself did not introduce noise. In short, the lower the noise figure the better.

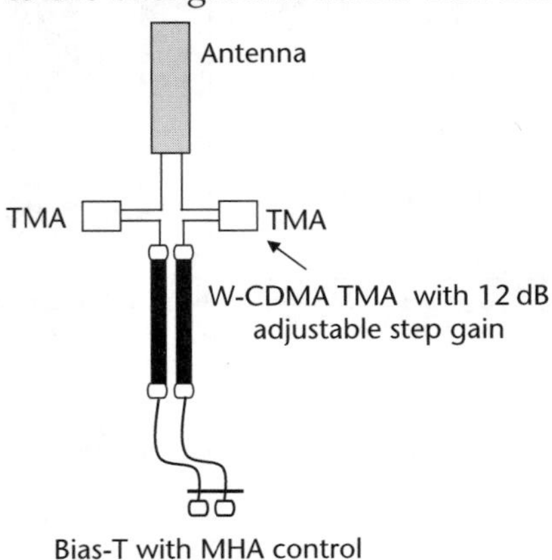

Figure 12.4 Tower mounted amplifiers (TMAs – also known as MHAs).

However a high-quality LNA should be able to offer a noise floor (NF) of around 1 dB with power gains up to 20 dB, or possibly even higher. Use of TMAs in large six-sector cells in a dense urban environment is not recommended, as it can reduce the coverage. This is due to the fact that wideband-code division multiple access (W-CDMA) modulation spreads the signal finely across the allocated 5 MHz band, hence, the uplink signal in particular is very close to the physical noise limit. Intercell interference is more prevalent within UMTS as it is more carrier/interference (C/I) limited than GSM, so the fact that the TMAs will also amplify the unwanted noise, as well as the wanted signal can cause a reduction in both capacity and coverage. In addition, TMAs are not recommended if employing the smart radio concept described earlier as again it is expected that this will also cause a reduction in coverage. Figure 12.5 represents the results of a case study performed throughout Europe for 80 sites assuming a data throughput of 144 kbps as compared to existing GSM sites. As can be seen, the six-sector site employing MHAs achieves less coverage than the six-sector site without any MHAs.

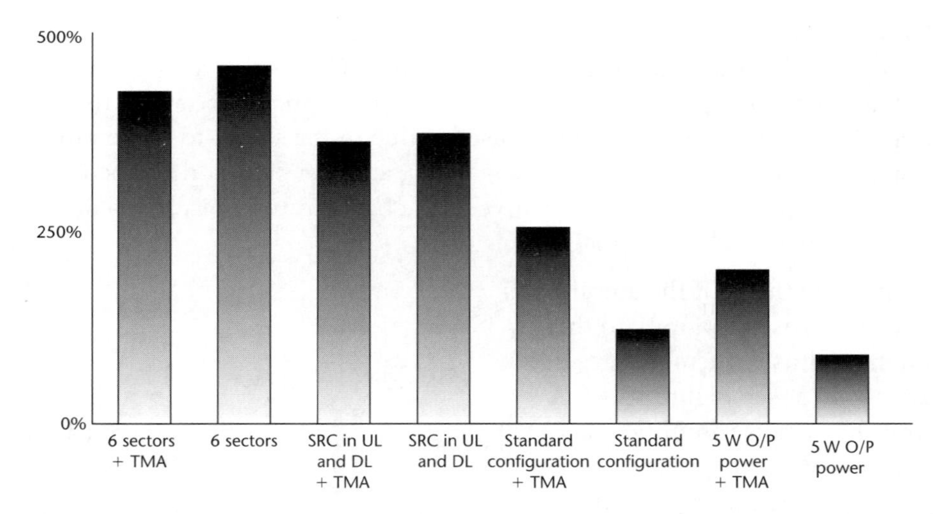

Figure 12.5 Various configuration coverage comparisons based on a case study of 80 sites for an urban area in Europe with 144 kbps compared to the existing 2G sites.

In summary, TMAs are advantageous in uplink-limited situations and are disadvantageous in capacity-limited dense suburban small cells. However it is likely that TMAs will be employed throughout most UMTS networks.

In the future this scenario may change, as increases in both traffic and subscribers will lead to cells being more densely packed within certain heavily populated city environments.

12.4 Antenna and Site Configurations

As a starting point, with regard to the decision-making process for selecting the correct antenna type and its configuration on the site, it is recommended to use a checklist as shown below, to ensure the numerous antenna mounting issues are correctly addressed:

a) Number and types of antenna due to be mounted. This may have an effect on both coverage and capacity dependent on the environment.
b) Obstructions that may affect the required coverage.
c) Receive antenna diversity and spacing requirements. Incorrect antenna spacing may cause a reduction in coverage.
d) Maximum cable run allowed. Cable lengths greater than the manufacturer's recommended distances will reduce the RF power, thus causing a reduction in coverage.
e) Isolation from other services/operators. Incorrect isolation will result in interference, leading to a reduction in coverage and poor QoS; in addition, it may also lead to dropped connections.
f) Antenna above-ground level (AGL) requirements. Incorrect antenna heights can cause coverage reductions and capacity reductions in adjacent cells.
g) Intermodulation requirements. Intermodulation can cause unnecessary interference affecting both coverage and QoS.
h) Antenna mounting parameters. Incorrect antenna mountings can affect the RF propagation pattern, and in turn affect the coverage.
i) Transmission path clearance requirements. Sufficient antenna heights should be used in order to ensure microwave path clearance is adequate. Inadequate microwave path clearance would impede the back-haul connection resulting in no coverage for the particular site.

Following the guidelines noted above the configurations of antennas noted below can then be chosen for deployment.

12.4.1 Three- and Six-Sector Configuration

A three-sector site would be implemented in a similar manner to a GSM standard site. However, with UMTS the most advantageous method would consist of a '1 + 1 + 1' carrier configuration, thus enabling the overall capacity to be increased. This three-sector configuration would typically comprise either 65° or 90° antennas which could be either standard or polarized. Another typical sectored site would consist of six sectors, when greater coverage and capacity is required. This configuration would employ polarization diversity antennas of 45°. Greater coverage can be realized due to a narrow beam antenna pattern and with the increase in antenna gain a capacity gain of around 80 per cent can also be achieved. However, in this type of configuration the antenna separation issues need to be taken into account, with regard to aesthetics, physical space available, intermodulation and isolation issues, etc., and these may cause difficulties in actual site design and implementation.

12.4.2 Omni-transmit and Omni-receive

An example of both omni-transmit and omni-receive (OTOR) and polarization diversity antennas are illustrated in Figure 12.6.

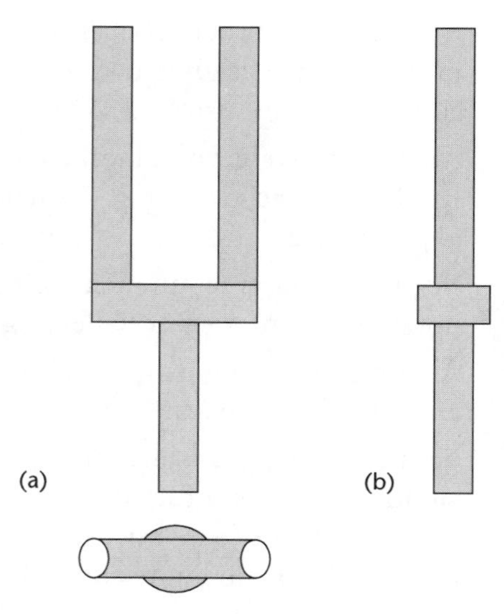

Figure 12.6 OTOR. (a) Standard and (b) polarization diversity antennas.

An omni-antenna has the ability to transmit in a 360° pattern, as opposed to directional antennas which concentrate the RF in a specific direction. With the OTOR configuration, both transmission and reception occurs in an omni-directional manner. This configuration can be useful in certain types of environments, however generally an omni-transmit and sector receive (OTSR) configuration (as illustrated in Figure 12.7), and is considered a more advantageous solution. This is because

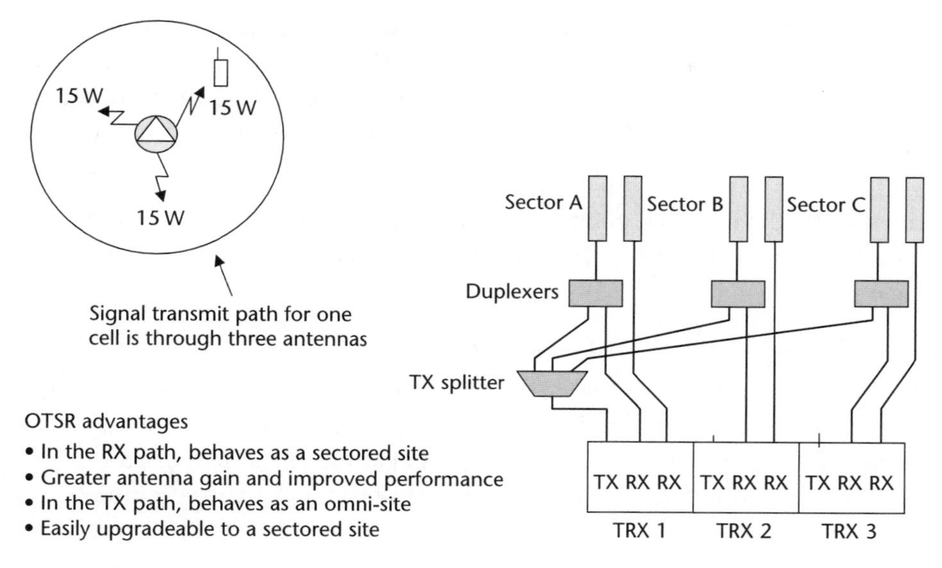

Figure 12.7 OTSR.

implementing this type of omni/sectored configuration, instead of a standard omni, which will ensure that only one linear power amplifier will be required in the base station. The OTSR configuration can be thought of as approximating to a standard omni-site with regard to the TX path, and in turn acts as a sectored site in the receive path. This is advantageous as compared to the OTOR configuration as it results in achieving a higher antenna gain, and thus better performance. Further benefits are the additional cost savings that can be made, due to the fact that linear power amplifiers are somewhat expensive. Hence this solution will assist in increasing both coverage and capacity, in addition to cost savings. Finally, this configuration also allows for easier upgrades to a full sector site when capacity requirements increase.

12.5 Site Optimization

This section covers the optimization techniques and issues that pertain to the radio planner. Areas such as uplink and downlink coverage gains, isolation requirements, co-siting antennas, and antenna tilting are covered.

With regard to the overall picture, there exist vast quantities of data which can be acquired from the radio network. The skill is to be able to filter and reduce this data to a workable level, extracting the important information which can then be used to make the decisions required to improve the network performance.

In essence, the network planning itself needs to be optimized to cope with both the increased traffic demand and spectrum availability.

12.5.1 Uplink Coverage Gain

As illustrated in Figure 12.8, a 2.5–3 dB gain can be achieved using receive antenna diversity, even if the antenna branches have fully correlated fading. This is due to the fact that the signals from the antennas can be combined coherently, while the receiver thermal noise is combined non-coherently.

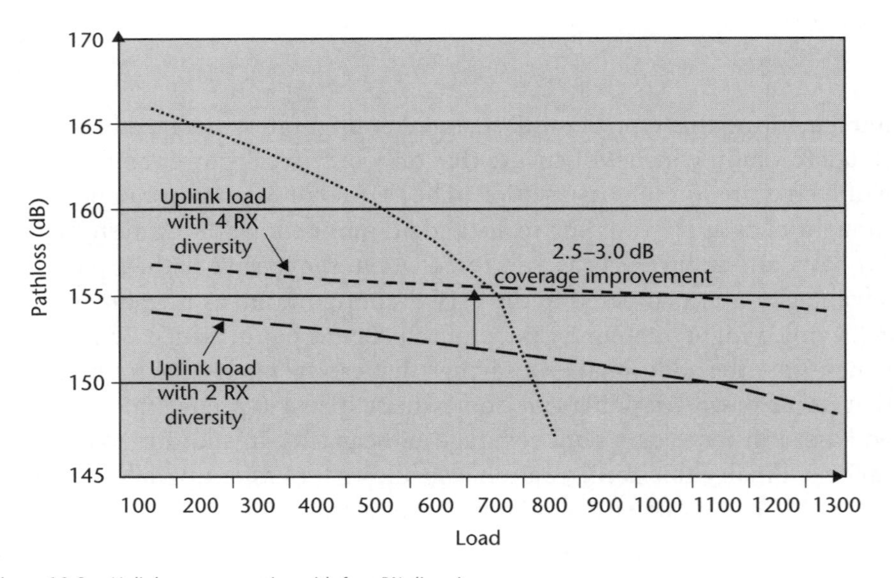

Figure 12.8 Uplink coverage gains with four RX diversity.

The 3 dB gain assumes ideal channel estimation in the coherent combining. Coherent combining can be thought of as adaptively obtained channel estimates, which are acquired in a manner whereby the impact of estimation errors on data detection is minimized.

This 3 dB gain is achieved because there are more receiver branches collecting energy, at the expense, however, of extra hardware being required in the base station. In addition, antenna diversity provides gain against fast fading, as fast fading generally does not correlate efficiently between diversity antennas. As previously described in Section 12.2, diversity can be achieved by space or polarization diversity. An advantage of polarization diversity is that the diversity branches do not need separation, and can be located in one physical antenna housing. The advantages here consist of achieving an increased gain, at the disadvantageous expense of installing extra hardware in the base station. However, dependent on the available budget, this is considered advantageous or even a requirement due to the increase in coverage that can be achieved.

12.5.2 Downlink Coverage Gain

It can be stated that as more power can be allocated per connection in the downlink than in the uplink, better coverage for high rate data services can be achieved in the downlink than the uplink. Assuming the planned cell is also required to possess the capability to bypass high data rates in the uplink from the periphery of that cell, then the cell will be smaller, and hence the downlink coverage will be better. However it should be noted that any loading of the neighbouring cells will affect the possibility of maintaining a high data rate connection at the periphery of the cell. An improvement in coverage can be achieved with a high antenna gain, and by increasing the number of sectors, and reducing the horizontal antenna pattern.

In the downlink, de-correlation between the transmit channels can be exploited by combining adaptive antennas with transmit diversity as described earlier in Section 12.2, as this will provide additional downlink capacity and coverage. It is also possible to configure the antennas at the base station to maximize the benefits from both adaptive antennas and receive and transmit diversity. To summarize: the downlink power is the limiting factor with respect to both coverage and capacity.

12.5.3 Isolation Requirements

Isolation can be described as the way an antenna interacts with its surroundings. The isolation will be higher the less the antenna is disturbed by its surroundings. With regard to multiple antennas, by measuring the energy directly transmitted from one antenna to another, the level of isolation can be determined.

The easiest and least expensive method of isolation is to mount the antennas some distance from each other. Since most installations are on a common tower, the antennas are normally mounted so that they are vertically separated from each other. Various solutions exist for achieving isolation, which can be achieved as illustrated in Figure 12.9. Firstly, by antenna selection and positioning; secondly, by filtering out the interfering

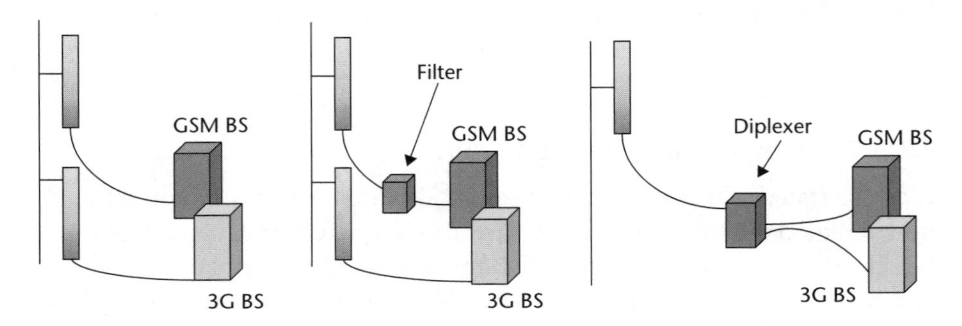

Figure 12.9 Isolation requirements.

signal; and finally, by using diplexers and triplexers with shared feeders and multiband antennas. Actual distances for antenna positioning cannot be given, as these will vary dependent on the antenna(s) and location, however the required information should be available within the manufacturer's specifications. Typical blocking isolation requirements between UMTS to UMTS and GSM 900/1800 to UMTS are approximately between 40 and 60 dB. Spurious emissions can be defined as emissions which are caused by unwanted transmitter effects (such as harmonics emission, parasitic emission, intermodulation products, and frequency conversion products). These can be measured at the RF output port of the base station.

Spurious emissions and intermodulation products can require in certain cases isolation of around 90 dB for some of the pre-99 GSM equipment. The isolation requirements will have a bearing on the type of antenna configuration and the filtering at both GSM and UMTS sites. This is advantageous when co-siting antennas.

12.5.3.1 Co-siting Antennas

It is inevitable that antenna co-siting will be a strategy pursued by most operators, mostly due to financial and environmental issues. However, there are some technical issues described below that the planner should be aware of.

Firstly, it can be difficult to calculate the isolation between two antennas, and therefore measurements must be taken. The best configurations will be where there is vertical separation between the antennas. This is because there will be less interference experienced by each antenna as there will be only minimal interference from the antenna's vertical lobes as opposed to the side lobes that are radiating in more of a horizontal direction, and hence are likely to interfere with each other. It can be stated that the type of antennas that are designed to suppress the side lobes will possess a wider beam width. This is explained in further detail in the next section (Section 12.5.4, entitled Antenna Tilts).

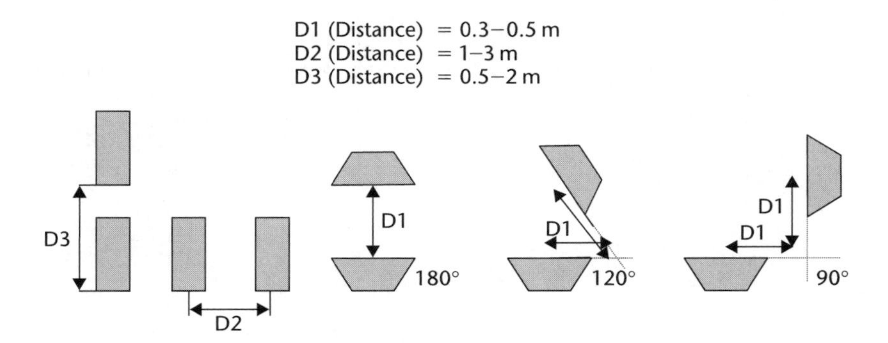

Figure 12.10 Co-siting antennas. These configurations are designed to provide an isolation of 30 dB.

The configurations illustrated in Figure 12.10 should give 30 dB of isolation; this can be considered the optimal figure and the most practical due to site limitations. The greater the isolation the better, but realistically higher figures can rarely be achieved due to site limitations and locational problems. When site sharing with third-party systems or other operators, again a minimum antenna separation between systems must be defined as some of the GSM systems may be more susceptible to electromagnetic interference (EMI) issues.

12.5.4 Antenna Tilts

Antenna tilting will have an effect on coverage and capacity levels, handover, and interference, hence it is advantageous for the planner to be able to optimize the RF patterns efficiently by the use of antenna tilting.

Antenna tilting can be used to control the actual antenna patterns. In addition regulating the RF power will also have an effect on the propagating pattern. The tilting of the antenna can be controlled by both the electrical and mechanical attributes and it is considered advantageous to use a combination of both. As antenna mounting is critical in the success or failure of a cell site, this aspect requires even more care for UMTS as strict control over the antenna patterns is imperative. This is the key to controlling the RF propagation within the required coverage area, as can be seen in Figure 12.11. Down-tilting the beam to point below the horizon,

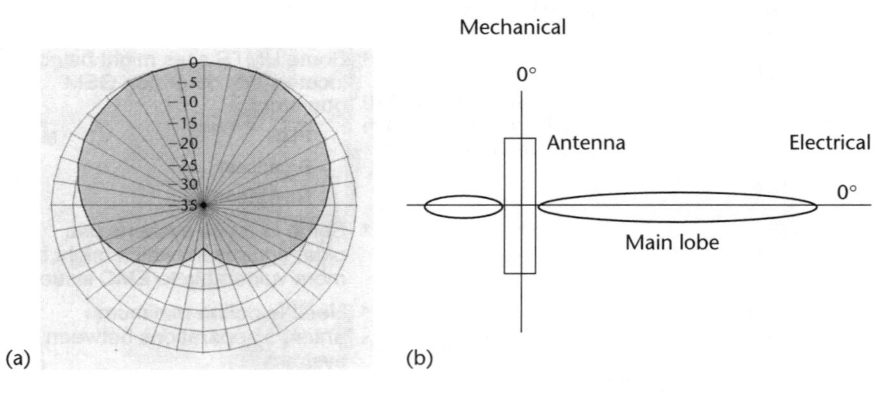

Figure 12.11 Antenna tilt Example 1: (a) RF propagation plot and (b) 0° mechanical and 0° electrical.

will cause the energy associated with the antenna beam to be directed into the required cell, and hence away from any adjacent cells. The signal strength in the desired cell will increase, and the interference will be minimized in the adjacent cells. The disadvantage is that a reduction in the coverage area is likely to occur, and the side lobes may increase.

Mechanical tilting can be easily understood as this involves physically tilting the antennas, and can have the effect of moving the rear and side lobes forwards, thus increasing the propagation pattern, which in turn could cause interference.

Electrical tilting can be best thought of as employing time delays along the propagation path of the signal feeds to the antenna. These time delays will produce variable phase shifts, which in turn will cause the output of the antenna to down-tilt the radiation pattern. One of the differences compared to mechanical tilting is that this will cause a downward tilt in both the rear and side lobes.

12.5.5 Mounting of Antennas

As both types of antenna tilting have such an effect on the performance of the base station, it is critical that the antenna mounting is both flexible and efficient in terms of implementation and operational needs for long-term optimization.

As antenna mounting is critical in the success or failure of a cell site, and requires even more care for UMTS configuration, strict control over the antenna patterns is imperative. This is the key to controlling the RF propagation within the required coverage area (see Figure 12.11). Below is a checklist which can be used as a starting point to ensure the issues regarding antenna mounting are correctly addressed:

a) Number and types of antenna due to be mounted.
b) Obstructions that may affect the required coverage.
c) Receive antenna diversity and spacing requirements.
d) Maximum cable run allowed.
e) Isolation from other services/operators.
f) Antenna AGL requirements.
g) Intermodulation requirements.
h) Antenna mounting parameters.
i) Path clearance requirements.

With regard to installing antennas onto a tower the physical spacing, or in other words, the actual offset (distance) from the tower must be calculated to ensure that the tower structure either enhances or does not affect the required antenna coverage pattern.

As illustrated in Figure 12.12, by increasing the mechanical down-tilt, the propagation pattern shows a forward reduction but the sides remain virtually the same, which can be beneficial in certain environments. When

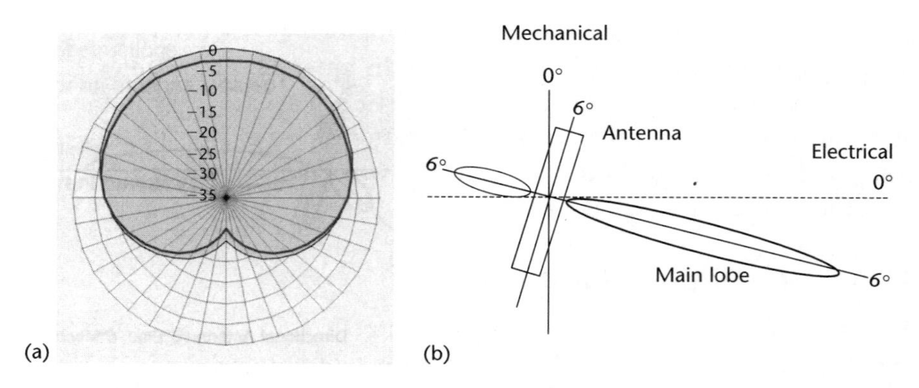

Figure 12.12 Antenna tilt Example 2: (a) RF propagation plot and (b) 6° mechanical and 0° electrical.

the coverage radius of a service area is set to a specific value, the higher the antenna is located, and therefore the larger the down-tilt angle, the larger the reduction in co-channel interference. Co-channel interference can be described as interference resulting from two or more simultaneous transmissions occurring on the same channel.

Co-channel interference is detrimental and can increase interference leading to reduced coverage and QoS. Increasing the electrical down-tilt in this example illustrated in Figure 12.13 gives a uniform overall reduction in the propagation pattern. This can be advantageous for densely packed

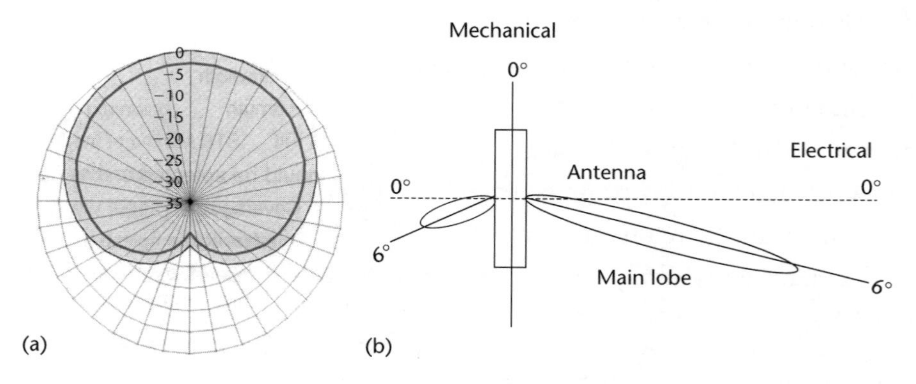

Figure 12.13 Antenna tilt Example 3: (a) RF propagation plot and (b) 0° mechanical and 6° electrical.

cells, whereby the RF emissions need to be uniformly reduced, to ensure no interference will occur in the neighbouring cells. The illustration in

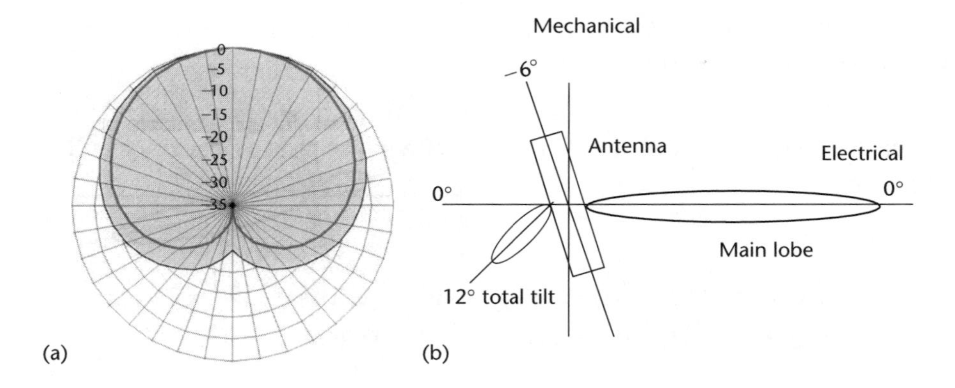

Figure 12.14 Antenna tilt Example 4: (a) RF propagation plot and (b) 6° mechanical and 6° electrical.

Figure 12.14 shows a combination of both electrical and mechanical tilting which causes a reverse propagation pattern reduction, while simultaneously causing a pattern reduction with both sides diminishing further towards the rear of the lobe. Antenna orientation must also be considered, as ideally to have no errors in the orientation of an antenna is desirable but somewhat impractical. The RF planning department should specify an orientation tolerance. However, if no specified tolerance exists, a general guideline or industry benchmark states that the tolerance should be within 5 per cent either way of the antenna's horizontal pattern. The tolerance for the orientation is reduced when there is a need for a tighter antenna pattern. This can be best demonstrated in environments that possess densely packed cells, as more accurate RF patterns are required in order to minimize unwanted interference. The objective should be ±5 per cent; however, this can be tightened or relaxed dependent on the system requirements. Building sway should also be taken into consideration, as this can easily occur, dependent on the type and height of the structure in question, and hence a maximum of 5 per cent tolerance should be factored in here to compensate for any movement. To summarize, as the network matures and becomes denser, and the amount of subscribers increases, tilt settings are likely to require further adjustment, so that new sites do not interfere with old ones. A study performed by Nortel Networks predicts that 60 per cent of deployed antennas will require re-tilting at least once to optimize capacity, and a further 50 per cent will require further adjustment as the network matures further and more sites are added.

Efficient tilting must be performed throughout the network as explained above; this will ultimately have a distinct effect on both the coverage and capacity.

12.6 Coverage, Capacity, and Quality of Service Versus Optimization Summary

Optimization is a key process as it is critical to mitigate the numerous risks involved with planning a network, and therefore it is important that the planner fully understands the key issues involved. Optimization covers many areas ranging from coverage and interference analysis to identify problem areas, drive testing for evaluating system performance, traffic analysis and recommendations to the optimization of system parameters, to mention but a few. However, this chapter covers the key areas that pertain particularly to the planner, from the antenna types, configurations and tilting to co-siting and isolation requirements.

In summary this chapter covers optimization as the final part of the CCQ model. The planner should always be aware of, and have the CCQ model in mind with regard to all issues with UMTS planning, as has been covered in the previous chapters (Chapters 3 and 4 relating to capacity, Chapters 5 and 6 relating to coverage, and finally Chapter 7 relating to QoS). All these fundamental issues interrelate with each other and are all affected by the issues covered in this chapter relating to optimization. As has been previously mentioned it is critical for the planner to continually take optimization and the various methods available into account, thus enabling the ultimate balance between these three parameters to be achieved in the most efficient manner.

Optimization, however, does not end here; planners of mature and developed networks must also consider the long-term future and therefore industry developments and changes. The following chapters, therefore, draw the attention of the planner to the integration of new technology and the potential for upgrades into their networks.

PART SIX

UMTS (3G) Development

CHAPTER THIRTEEN

UMTS (3G) Development and Future Technologies

The mobile industry is continuously evolving and changing. This chapter covers some of the new developments already existing along with some of the possible future technologies. In addition, the business case must also be considered, as today's operators are increasingly nervous and wary of which path will prove successful and generate those much needed revenues. The basis for all future technologies is driven by the ever-increasing demand for more coverage, capacity, and finally the ultimate quality of service (QoS).

13.1 Main Components of 3G Universal Mobile Telephony System Services

Universal mobile telephony system (UMTS) services will consist of numerous components that will ultimately provide a variety of information to the end user, such as Internet service providers, content providers, mobile, and virtual mobile network operators. Numerous methods will emerge for individual users to access, manage, and control vast amounts of information from many diverse sources. The UMTS Forum (www.umts-forum.org) has already identified various key revenue components such as customized infotainment, mobile Intranet/Extranet access and multimedia messaging services. Further key features or products are seen as location-based services, gaming applications, M-commerce, and advertising and sponsoring.

The availability of attractive user applications and devices must also be considered, as it is predicted that various applications will require larger screens, fewer keys, and more functionality. Personal digital assistants (PDA) have the potential for use with gaming applications as well as in the business arena. For these services it is important to ensure that timely content delivery and predicting realistic expectations for the end user can be realized.

With third-generation (3G) the emphasis has always been on finding the key revenue generating application, however it is apparent that a 'cocktail' of killer applications will prevail. For certain applications sensitivity to differing cultural and local demands will be required. Simultaneously, there will be similar services emerging across the world, so it is likely that a group of applications will become a source of high revenue.

As has been seen with second-generation (2G) voice communications, their ability to generate profits is proven, however that will change with UMTS. Operators will require these new revenue streams with UMTS (3G) that will be derived from data access and movement.

The evolution of global system for mobile communications (GSM) cellular communications began with operators focusing on high spending business users, followed by capturing the mass market to increase user penetration to achieve close to saturation of the working population as quickly as possible. A full circle has now been completed, as today operators are turning their attention back towards increasing their high revenue generating customers. The pressure to do this will be quite intense, especially when considering the high investments in 3G licenses made by some operators. It is likely that a similar sum will be required again to build, implement and roll-out the network.

The potential key applications will ultimately determine the success of 3G. Wireless 'on-line' gaming and video streaming are seen as the possible popular applications, however, these will still require further improvements in the UE terminal battery life.

Location-based services will be the catalyst for many future new services, so much so that the Location Interoperability Forum (LIF) has been formed, which will enable simple, secure and interoperable services globally. Some of the location-based services that will be possible will provide such services as geographical location tracking, navigation, local map information, and city guides.

Finally, mobile messaging has already proved to be a success and has achieved a considerable market share in current GSM networks. Short message service (SMS) has developed into a well-accepted value-added service and has surpassed all original expectations, and thus instills confidence in the messaging market of the future. Therefore, it is likely that mobile multimedia messaging (MMM) services, only possible with UMTS (3G), will also prove as popular. This upgraded version of the traditional SMS will allow the end user to be able to submit and receive messages with multimedia content, such as video, audio, and e-mail. Initially it is perceived that such services consisting of electronic postcards and unified messaging are both expected to be embedded within a personal portal. The major infrastructure vendors have now formed a MMM service interoperability group to ensure MMM deployment can be smoothly implemented by mobile service providers worldwide. Ultimately one of the key services will be the ability to access the Internet remotely either through the handset itself, or as a connection point for a laptop or PDA.

As this is a dynamic process, whatever services are currently being envisaged, there will be at least as many again which are as of yet unimaginable.

13.2 Universal Mobile Telephony System (3G) Business Case

The business case for some European operators would have been higher if the UMTS license fees had not reached such extreme levels due to the technological boom of the early 2000s, as the roll-outs could have been better structured. The UMTS forum forecasts a compound annual growth rate of over 100 per cent during the first decade of UMTS, with total revenues for the following three main predicted services, customized infotainment, intranet/extranet access and multimedia messaging, together are forcast to exceed $164 billion by 2010. Table 13.1 shows the investments made in the 3G, licenses for the five operators in the UK.

Table 13.1 3G Network building costs.

Country	Network	Licence cost (GBP) (Billion)
UK	Hutchison	4.38
	Vodafone (UK only)	5.96
	One2One (T-Mobile)	4.00
	mm02 (BT)	4.03
	Orange	4.09

Source: BBC, December 2001.

However, one advantage of the UMTS network, in its infancy, will be the possibility to increase revenue income from the ability to utilize the large bandwidth for additional low rate 2G voice calls, increasing the capacity for the real 3G services when required.

It appears that average revenue per user (ARPU) as a whole is declining, but this is mainly an effect of the markets beginning to reach saturation with more of the low volume users. In reality, the heavy business users are still high consumers of the capacity, but they are now sharing the networks with far more infrequent users resulting in an overall fall in voice revenues. Operators will have to supplement their income from data services that are yet unproven, coupled with ensuring that projected subscriber figures are achieved. Different challenges await new entrants as they will require an ARPU higher than today's equivalent. One of the strongest effects on an operator's business case is the population density which will have a knock-on effect on the network as a whole, especially in the early stages of the development. This is a catch-22 situation for the operator as subscriber levels allow for network improvements, but subscribers will not join a network if the coverage, capacity, and QoS (CCQ) elements are weak.

Another interesting example to be considered is that of an existing operator based in a country with a population density such as Sweden. For all intent and purposes, assuming this operator has invested a similar sum in the license fee, this operator would have to achieve a higher ARPU than that required by the previous existing operator, based in a highly, densely populated country such as Germany.

Today's operators are already experiencing financial difficulties, and this will force them to optimize their business cases. Taking a more optimistic view, it is generally expected that based on predicted future demand for the wide range of mobile services soon to become available, a higher ARPU will be achievable. It will be possible for existing operators to survive as well as generating profits. However, in certain countries other issues must be considered. Restrictions are coming from the regulators who are imposing strict requirements, forcing the operators to achieve high QoS levels, unrealistic coverage levels, and national roaming within an unrealistic specified time frame. Due to the competition for the licence, assuming some of these requirements are not relaxed, this will threaten the business case and lead to increased costs and penalties. The key competitive advantage for an operator is being able to offer extensive coverage. Ideally population coverage should be as close as possible to 100 per cent within urban areas, at least in countries with high population densities. As QoS levels improve, additional coverage requirements can be implemented. One of the problems being experienced in the development of a UMTS network in the late stage of 2002 and the early part of 2003 was the availability of 3G handsets. As has already become evident, the limited numbers of handsets available so far have been plagued with software problems, and subsequently this will have a critical effect on both the usage and acceptance of future 3G services.

One of the greatest impacts on the business case concerns the level of handset subsidies. Within the marketing and sales budget allocated, this is the highest cost item and the most unpredictable.

Finally, alternative technologies must be considered. It appears that the only credible competitor within the high data rate wireless arena is considered to be wireless local area networks (WLANs). WLANs are based on the IEEE 802.11x standard, and are implemented as an extension to wired LANs within a building. WLANs are already implemented in many libraries, cafes, and strategic locations within certain cities throughout Europe. The mobile functionality only exists over the final few meters of connectivity between the wired network and the mobile user, consequently the range is very limited. This distance will vary dependent on the type of technology and location. However, at this stage it is seen to be an unlikely threat to 3G. Coverage with WLANs is very limited and hence will require a completely different business model. It is a distinct possibility that this technology may even complement 3G services. In such a case WLANs may consume some 3G traffic, but with such limited coverage ability, it is not seen as major risk that could threaten the success of UMTS.

13.3 Implementation Aspects

The implementation of the actual network will mainly be threatened by the lack of access or planning permissions for the base stations. However, as already seen today, the majority of the problems occurring are handset related problems, including the actual lack of a choice of reliable handsets currently available on the market. Multimode terminals which will be both 2G and 3G compatible will only exacerbate the implementation problem further, particularly as battery technology is not yet advanced enough to provide adequate power for the ever power hungry UMTS handsets. This is similar to the problem that occurred when the very first mobile networks were introduced, but as the handsets then were mainly restricted to cars the battery problem was not an issue, this will not be possible in a 3G network. Ultimately a solution will be found, whether this is from improved batteries or from fuel cell technology. But in the interim the lack of a consistent power supply could jeopardize the operator's offer and promise of all the new functions. With regard to the 3G partnership project (3GPP) standards (see www.3gpp.org), operators need to be very careful with their deployment strategy and determine whether to deploy the 3GPP release 99, as seen and explained in Chapter 1, in an attempt to initially gain a larger portion of the market share. Alternatively, it may be wise for certain operators to wait, and implement the future 3GPP 2000 releases 4 and 5, as this will ultimately take care of their future evolution path, while protecting their investments and reducing future CAPEX costs.

In summary, the need for both mobility and information access is likely to increase and hence the requirement for widespread access to information is still driving UMTS forward, albeit at somewhat of a slow pace at present. However, with the current market slowdown, coupled with the technical difficulties being experienced, it is difficult to predict which technology will eventually become the dominant force with regard to high speed data transmission. This has become apparent with the ever-increasing WLAN access points now being installed in numerous cities (refer to Chapter 14, entitled Wireless Technologies of the Future), along with the current interest in implementing EDGE technology into the existing 2G networks as a means of offering high data rate usage at an affordable price. For UMTS to survive it is critical for both the vendors and operators to be able to resolve the technical issues in a timely manner and to be able to offer these long awaited and promised services at an affordable price to the end user.

Wireless Technologies of the Future

14.1 The Move towards 4G

As with most technologies, no sooner does one come into fruition, than all the attention switches to the next generation. In the wireless arena, the future generation, in this case the 'fourth generation' (4G), will be required to provide higher speed, greater capacity, be able to offer Internet protocol (IP)-based services at an affordable price, and finally support multiple access technologies including the next stage towards automated processes.

This next phase will signify a fundamental change in the telecommunications industry, as the emphasis will be based on data movement. Previously, both data and voice communications were detached from each other; however, in the communications systems of the future, there will be no difference.

14.2 Stable 2G and 3G Networks

The current second-generation global system for mobile communications (GSM 2G) networks, in the majority of countries, can now be considered as mature due to the extensive network of base stations providing virtual total coverage (by population). However, with universal mobile telephony system (UMTS), the concentrations of base stations especially microcells will be far greater, in order to ensure adequate coverage and capacity can be achieved in the urban environments. As is well known today, site acquisition is likely to prove very challenging due to the resistance by local authorities in view of the alleged health issues associated with radio frequency (RF). Stability of the third-generation (3G) networks can only be achieved, therefore, through the careful continual monitoring and optimization of due to the as yet unknown traffic both currently in the network and in the future.

Initially, UMTS 3G coverage will extend throughout the major densely populated urban areas within large cities and will expand outwards to suburban regions and major highways. It is inevitable that further expansion will occur, however this will be a cost–benefit decision rather than the need to provide total coverage.

14.3 Market Developments and Demands

As most cellular networks will be based on IP, it is likely that equipment providing dedicated cellular activities will exist with respect to the type of service required by the end user. These will be call or connection processing servers. The ultimate aim will be a merged voice/data/multimedia network, with the ability to achieve high user capacity and high data speeds. It is likely that the demand for multimedia coverage will push the requirements and therefore the demand for coverage and capacity. This will most likely be in the form of broadband wireless video and will encompass services, ranging from video on demand, sport coverage, or entertainment guides, which may consist of movie clips, to interactive business guides and highway travel information, including real-time web cameras showing traffic hot-spots, real-time video briefings and breaking business news.

The benefit of using all IP is that the core network will be able to evolve independently from the radio access network, as IP will accept a wide range of protocols, some of which have not yet been developed. 'Standards' organizations must also play their part in working to advance international acceptance of what will essentially be 4G protocols (a brief description of 4G can be found in the following section).

Along with the development of 3G/4G, the merging of wireless local area networks (WLANs) technologies can be thought of as a natural progression in the never-ending quest to provide seamless higher and faster data transfer using WLANs (WLANs are covered in more detail later in this chapter). These networks will allow UEs to be able to utilize high-speed 11-Mbps data streams (802.11b) and in the future 54-Mbps data streams (802.11a and g), when located within certain corporate or hot-spot areas. In short, it is likely that WLANs will be seen to complement UMTS (3G) and can be thought of as the next stage in the evolutionary path.

14.4 4G

The fourth generation can be best thought of as a combination of an advanced air-interface system using a superior modulation scheme, thus enabling higher and faster data transfer over the air interface, coupled with an all IP scenario within the core of the network (reviewed later in this chapter). A number of evolutionary changes have occurred throughout the previous radio networks, as can be seen in Figure 14.1, however more will

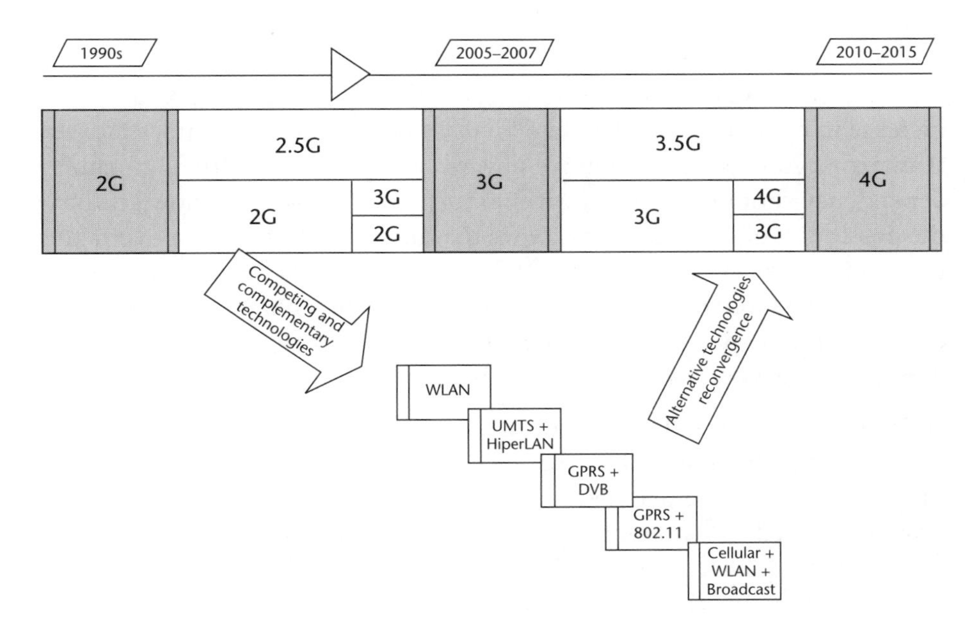

Figure 14.1 Future developments in broadband wireless communications.

yet be required, although the majority of future changes will take place within the operation of the core network and not at the user end. Over time the quality-of-service (QoS) issues, with regard to IP and mobility, will be fully resolved and IP will eventually be able to support these new developments. Even though we are not ready for any 4G planning and testing, research and development must continue in this area if carriers are to be ready to handle the expected long-term large demand for high-speed wireless data services. It is likely that IP-based 4G networks will provide data throughput speeds from 20–100 Mbps. Such types of networks will need greater spectral efficiency, and a dormant to active transition time of 0.1 s.

14.5 The Move from 3G to 4G

Once UMTS (3G) has reached a stable platform, enhancements to the existing system can and will be made. The aim here is to provide improvements to the wideband-code division multiple access (W-CDMA) air interface along with improving spectral efficiency, ensure backward compatibility with W-CDMA, and optimize the resource control for IP. These improvement stages can in some ways be thought of as a development to 3.5G.

Enhancements within the air interface are likely to result in a six times increase in capacity and a data rate increase of around fifteen times more than expected with 3G. The move towards freeing the user from operator dependency should also begin within this phase. The stage is now set for the possibility of various technologies that may enhance or even overtake the original concept of 3G; this will, however, require a flexible integration of various air interfaces. An overview of wireless broadband technologies can be seen in Figure 14.2, and the relevant technologies shown in this illustration are covered throughout this chapter.

To introduce 4G, it will comprise a network that is able to support higher-data transmission rates than 3G. The 4G mobile data transmission rates are planned to be up to 20 Mbps. Although it is not possible to predict technology developments, a 4G system will in principle support high-quality smooth video transmission. In addition, a wide range of access and networking capabilities will ensue. It is widely believed that 4G will consist of an IP-based network.

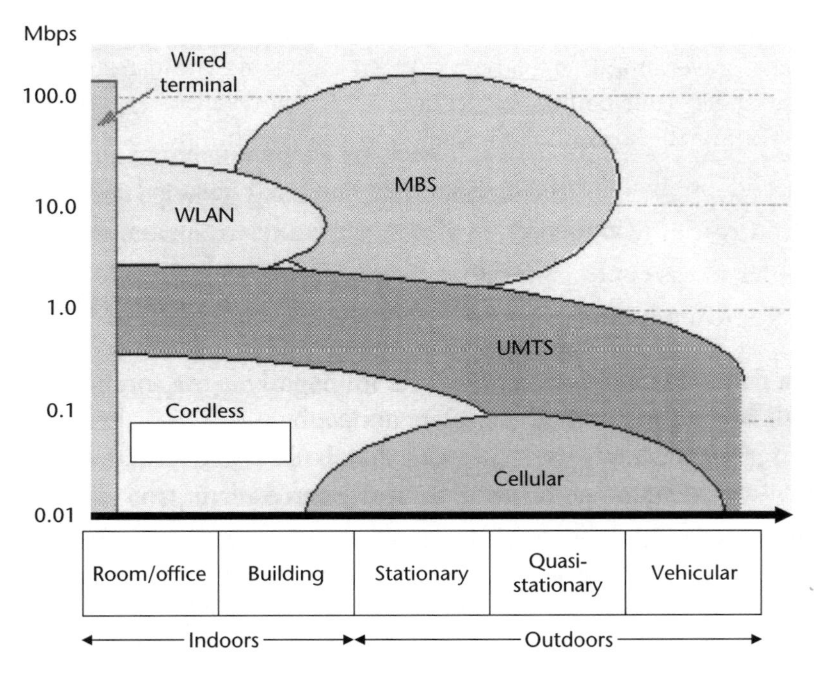

Figure 14.2 Bit rates versus mobility.

14.6 Technology Required for 4G

As most operators continue to battle with debt, falling share prices (resulting in less equity for expansion) and roll out delays with regard to UMTS, nevertheless, the larger network providers and wireless carriers are already investing in research into the next generation of mobile communications, namely 4G. Hence, to remain competitive, research and development into future technologies must be considered a perpetual, ongoing process.

It is likely that several 4G concepts will emerge during the ongoing 3G implementation and will thus be seen to become instrumental in future 3GPP releases. Although today's operators are reluctant to see future wireless developments move ahead too fast at this point, as they are still recouping their previous investments, they are aware of the need to remain at the forefront of technology to ensure future survival in this constantly changing industry.

The ongoing research and development will result in improved wideband receivers, power amplifiers, advanced faster modems, multi-beam antenna systems, space–time coding, multi-user detection (MUD), and RF transceivers. But further research will also be required within the following areas:

a) Fast adaption algorithms;
b) Multi-carrier technology;
c) Media access control algorithms;
d) Multiple input, multiple output (MIMO) algorithms;
e) MIMO antennas for both terminals and base stations;
f) MIMO propagation and channel modelling;
g) System simulation;
h) Reduced complexity receivers.

One of the major areas to be addressed is the improvement required in radio technologies. This is due to the limited data rate transfer possible via the current rather poor modulation schemes being used. At present, there appears to be a lack of formal methodology with regard to software radio implementation on reconfigurable platforms. One possible method currently under study involves a layered radio architecture using stream-based computing, which would enable software updates and validation to be performed across the air interface. Power amplification design and implementation are likely to prove costly in the short term. It is possible that power amplifiers could end up accounting for a large percentage of the total costs of a 4G base station due to the complex design. Being able to compensate for non-linearities within power amplification is a fundamental pre-requisite in advanced power amplifier design. Digital pre-distortion would reduce non-linearities, which cause high levels of power dissipation, and thus only add to poor system performance. Digital pre-distortion is achieved by accurate modelling of laterally diffused metal oxide semiconductors (LDMOS) power devices. LDMOS devices provide better thermal characteristics, hence giving a greater reduction in thermal noise. Using digital correction techniques can also enhance efficiency as well as help to reduce costs.

In summary, a user-driven approach is likely to influence 4G development, along with IP convergence, reconfigurability of both the radio layer and applications, and finally, it is likely to include a hierarchy of wireless networks, for example WLANs and mobile broadband systems (MBS) covered later in this chapter.

14.6.1 Wireless Technologies

WLANs and MBS technologies can be considered as the basis for the 4G. The WLAN technology is already being implemented and is at present functioning in various cities throughout the world. MBS, although under development, will emerge in the not too distant future and should be able to handle data rates up to 160 Mbps. This technology can be thought of as a cellular-type system, which will allow movement speeds exceeding that of WLANs. WLAN is a flexible data transfer system, and can utilize both RF or infrared (IR) technology. They are basically the next phase in the development of traditional local area networks (LANs), which included technologies such as Ethernet and Token Ring, but with the freedom and luxury of being a wireless system. These WLAN systems are based on the IEEE 802.11 standard, and will utilize a more efficient modulation scheme in the physical layer.

14.6.1.1 Wireless Local Area Networks

WLANs are considered to be a promising technology which could be used in the next step of the evolutionary path in the quest to achieve higher data speeds. Also to be taken into consideration are the IEEE 802.11x standards, which can be thought of as the leading standard for wireless networking (also known as WiFi), running on three channels in the unregulated spectrum at a frequency of 2.4 GHz. This part of the spectrum is also shared with Bluetooth products, cordless phones, and microwave ovens.

The 802.11b standard enables data throughputs of up to 11 Mbps in the 2.4 GHz range and the 802.11a standard will provide data throughput rates up to 54 Mbps in the 5 GHz range. At this stage it is still unclear what the relationship will be with these emerging technologies, but due to their specific usage (static versus mobility), we believe that in the long term they will complement each other. It is clear that UMTS (3G) will provide true mobility, whereas WLANs will only provide high-speed access to users in certain public areas and office environments. Regardless of the type of utilization of these systems, they are both being implemented together and, in the case of the WLAN systems, they are already found in many social or academic locations in many cities in Europe and the United

States. One of the key drivers for WLAN systems is the corporate adoption of virtual private networks (VPNs), whereby business users with notebooks are able to gain wireless access with relative ease within the office, home, and public areas, such as airports, convention centres, railway stations, cafes, etc. However, in the product life cycle, it may transpire that potential users will not require such high-speed access while on the move and will be satisfied with just being able to attain access in certain locations.

On the downside, there are security issues with the wired equivalent privacy (WEP) security protocol used in the 802.11b standard. It would be more beneficial to run Internet protocol security software (IPSec), as WEP will only protect the final link, whereas IPSec can be run end to end. Due to the requirement for increasingly complex high linear power amplifiers, multi-carrier modulation (MCM) was not adopted as standard for the European high-performance radio LAN (HIPERLAN). However, as deregulation has taken place within the industrial scientific and medical (ISM) bands, this has opened the door for manufacturers to develop their own wireless technologies. This has also allowed various companies to engage in research for more improved modulation methods, such as multi-carrier-CDMA (MC-CDMA) and orthogonal frequency division multiplexing (OFDM). OFDM is covered further in this chapter.

In addition, there are two new mobile air interfaces which have been specified by the IEEE 802.16e and 802.20 working groups. With regard to the 802.16e which is currently under development, this will provide mobility to stations that mostly support fixed wireless networking in the 2 GHz to 6 GHz range. The 802.20 was formed by the IEEE with regard to packet-based IP wireless networks. The thought behind 802.20 is to offer comparative data speeds usually offered by DSL services, ideally giving around 1 Mbps whilst the user is travelling at physical speeds of anywhere from 150 to 250 km/h. Hence it is perceived that when using this technology for the 'last mile' connection it would be able to support users in trains, and vehicles traveling at high speeds with an extremely low latency (under 20 ms or less). This is also still under development as of this publication going to press. One of the advantages of 802.20 is to optimize an IP friendly, mobile wireless interface taking into consideration today's packet-switching architecture, hence mobilising the PC. How this standard will function is still unclear today, however it is likely to be supported by using OFDM (Orthogonal Frequency Division Multiplexing)

(OFDM is covered in more detail in Section 14.7 of this chapter). However it is important for the reader to be aware that these three technologies consisting of 3G, 802.16 and 802.20 are likely to have huge potential, although it is unclear whether they will finally compete or complement each other at this stage.

14.6.1.2 Mobile Broadband Systems

MBS can be considered as part of the evolutionary process from UMTS (3G) and is likely to follow and complement WLANs. This progression is based on the need to satisfy the ever increasing demand for higher volume and faster data transmission. MBS will provide higher data throughput rates than UMTS (3G); however, the limitations are similar to the related WLAN systems, thus providing access for the stationary or quasi-stationary user. An example showing data rates against mobility is illustrated in Figure 14.2. The MBS will provide restricted radio coverage within the 40 GHz band, and is planned to support data rates of up to 155 Mbps. The driving force for MBS is the growing demand for mobility and the ever increasing popularity of laptop computers and personal digital assistants (PDAs).

Cost savings can be implemented with MBS as the wiring infrastructure is minimal, as these systems will only operate at the 'picocell' level. Research in MBS is currently underway within the advanced communication technologies and services (ACTS) programme, along with the definition of system aspects, network management issues, and the radio integration, all having yet to be resolved. MBS will support video mobile applications including compatibility with WLANs and wireless local loop (WLL) systems.

The system for advanced mobile broadband applications (SAMBA) project is a joint European project to promote the development of a broadband cellular radio, and develop a trial platform consisting of two mobile terminals and one base station operating in the 40 GHz band.

14.7 Orthogonal Frequency Division Multiplexing (OFDM)

The term orthogonal frequency division multiplexing (OFDM) refers specifically to the type of modulation scheme that will be used within the air interface. As the air interface is regarded as the weakest and most vulnerable link in the chain, it is critical that a robust method of radio transmission is used. OFDM can be considered as part of 4G, hence with this type of efficient modulation system, the remaining changes of network evolution towards an 'all IP' scenario, along with the new associated protocols that will be required, will ensure that a high-speed, high-volume data rate transfer 'path' can be achieved, and in doing so it will create 4G.

OFDM makes efficient use of the available radio spectrum, therefore offering more bandwidth per hertz. One of the benefits of OFDM is that it is very resilient to signal fading, which is an inherent problem with data transmission, especially when high data rates are being used. This is made possible as the data bits can be 'packed' onto numerous carriers, hence the individual transmitted signals can be of a longer duration. Consequently, this limits the possibilities of fading and interference. Therefore, OFDM will use multiple carriers or tones to divide the data over the radio spectrum available. The difference to a standard frequency division multiplexing scheme is that each tone can be thought of as being independent to the neighbouring tones, and will not need a guard band. The guard bands here will encompass a set of tones, thus improving the spectral efficiency. Alternatively, the transmitted data streams are divided into multiple parallel bit streams that will possess a much lower data rate than the original data rate. These multiple parallel bit streams will then be used to modulate several carriers. Similar to CDMA, OFDM is not a new phenomenon and has been around since the 1960s, but has only recently been considered for high-speed wireless data transfer. Initially, data rates of 56 Mbps should be possible; however, one of the major problems with OFDM is the power consumption required to generate this modulation scheme. Eventually, technological advances will overcome this and research is underway in software-based radios and wideband transceivers that will be capable of using OFDM. The initial phase will consist of software radio deployment in the base stations, followed by implementing smart antenna solutions that are OFDM compatible. OFDM has a greater resistance to multi-path effects compared to conventional single carrier modulation schemes and the data rate per sub-carrier is selected in such a way that any inter-symbol interference (ISI) can be avoided. The OFDM concept was developed in the late 1960s. However, by using fast Fourier transform (FFT) techniques and faster digital signal processors (DSPs), OFDM has become a popular modulation scheme and will be used in the new 5 GHz band comprising of the 802.11a, HiperLAN2, and WLAN standards. The high-performance radio local area network type 2 (HiperLAN2) consists of 5 GHz radio networking technology, designed for operating in LAN environments. HiperLAN2 possesses interconnection capabilities into almost any type of fixed network technology. Hence, it is suitable to connect mobiles, laptops, and PDAs to a fixed access point.

In addition, it is worth mentioning flash-OFDM, This has been developed by Flarion Inc. in the USA, and will be able to offer a fully mobile air interface for broadband connections. Although very similar, flash-OFDM utilises a more efficient and effective error coding correction scheme known as LDPC coding (Low Density Parity Check). Data throughput speeds of 1.5 Mbps and up to 3 Mbps will be achievable, reducing only to 384 kbps at the periphery of the cell.

Flash OFDM consists of an integrated physical and media access control layer, along with pure IP-based layers above the actual network layer (layer 3). Applications using standard IP protocols are supported by these IP-based layers and as well as the high broadband speeds, a very low latency is also achievable, thus enhancing the all important QoS, as well as offering full mobility and high spectral efficiency. When taking the current 3G air interface into consideration, which was originally designed to operate with both circuit-switched and packet-switched networks, they are just not able to offer such high speed, low latency data transfer possibilities, as when using such a technology as flash-OFDM. This coupled together with 802.20 could well possibly become a formidable competitor for that all important 'last mile' connection to the end user.

To conclude, OFDM is also known as a multi-carrier or discrete multi-tone modulation and is the modulation scheme currently in use for digital television in certain countries within Europe, Japan, and Australia. In addition, OFDM is also utilized for the digital audio broadcasting (DAB) standard within Europe, along with forming the basis for the global asymmetric digital subscriber line (ADSL) standard.

14.8 Future Vision of Communications

Eventually, 4G networks will deliver the ever increasing data rates, exceeding all current 3G expectations. Therefore, it is inevitable that the advances made within 4G will ensure high-speed data and voice over

Internet protocol (VoIP) becomes a reality. Moving further ahead along the communications road, it is difficult at this stage to accurately predict how future developments will evolve. It is however safe to assume that communications of the future will consist of devices becoming more and more capable of sensing and interacting with each other, and will also become more diverse, eliminating the need for the keypads of today, as the majority of requests will be performed by voice commands. There is a pattern emerging, whereby the convergence of computers, communication systems, micro-electrical–mechanical systems (MEMS), and sensor technologies will evolve into a global digital nervous system, hence fundamentally changing the way we live our lives. User terminals have evolved and will evolve further to the additional dimensions of communication as we know it by using more of the human senses of touch recognition, retinal scans, smell, and possibly taste. Services that today seem unimaginable will be commonplace enabling the end user to experience these senses. Air-interface development, always seen as the weakest link in the chain, will allow seamless, real-time connections with no restriction on data speeds.

In the not too distant future, automatic language translation will be possible, enabling an end user to communicate in his mother language, say, English, and be immediately understood by the other party in say, Russian or Chinese!

Potentially at the current development rate, we may see the full emergence of 4G around 2010 following which a completely new communications technology will emerge that is currently unheard of today, as it may not be based on any form of radio system, but is likely to have a closer link to the user interface. This itself could evolve into many different forms once the restriction of the keyboard has been removed. The rest is currently science fiction, but it is not unlikely that, regardless of location, individual access to the internet and other users will be constant with zero delay, and therefore will provide a seamless communication thread to what could then be known as the 'globalnet'. Basically, as we have seen over the last twenty years, from the early 1980's, that there is an emergence of communication media into everyday life. These will develop and mutate further again over the next twenty years, so that the network planner will either be redundant or, more likely, will have to be aware of more options, and will have to embrace these technologies in order to maintain a competitive network.

References and Bibliography

Chapter 1: Introduction

BWCS Mobile Data Daily; www.bwcs.com.

Holma, H., Toskala, A. and Latva-aho, M. (1998). Asynchronous wideband CDM for IMT-2000. *SK Telecom Journal*, South Korea, Vol. 8.

Nikula, E., Toskala, A., Dahlman, E., Girard, L. and Klein, A. (April 1998). FRAMES multiple Access for UMTS and IMT-2000. *IEEE Personal Communications Magazine*.

Prasad, R. (1996). *CDMA for Wireless Personal Communications*. Artech House.

Prasad, R. (1998). *Universal Wireless Personal Communications*. Artech House.

Westman, T. and Holma, H. (1997). CDMA system for UMTS high bit rate services. *Proceedings of VTC '97*, Phoenix, Arizona.

Part I: Network Planning and 3G Foundations

CEC Delivery No. R2020/TDE/PS/DS/P/040/b1 (June 1994). CODIT final propagation model.

Cheung, J.C.S., Beach, M.A. and McGeehan, J.P. (November 1994). Network planning for 3rd generation mobile systems. *IEEE Communictions Magazine*, Vol. 32.

Hartman, Lappetelainen and Holma (1997). A novel interface between link and system level simulations. *Proceedings of ACTS Mobile Communications*.

Holma, H. and Toskala, A. (2000). *WDCMA for UMTS. Radio Access for 3rd Generation Mobile Communications*, p. 11. John Wiley & Sons.

Jakes, W.C. (1974). *Microwave Mobile Communications*. John Wiley & Sons.

Ojanpera, T. and Prasad, R. (1998). *WCDMA for 3rd Generation Mobile Communications*. Artech House.

Pirzarroso, M. and Jimenez, J. (September 1997). Common basis for evaluation of ATDMA and CODIT downlink capacity enhancements. *Proceedings of PIMRC '97*, Helsinki, Finland.

Yang, S. (1998). *CDMA RF System Engineering*. Artech House.

Part II: Capacity and Network Planning

Andermo, P.G. (1992). System flexibility and its requirements on 3G mobile systems, *Proceedings of PIMRC*, Boston, Masschusetts, USA.

Dahlman, E. and Jamal, K. (May 1996). Wideband services in a DS-CDMA based FPLMTS system. *Proceedings of VTC '96*, Atlanta, Georgia, USA.

Faruque, S. (1996). *Cellular Mobile Systems Engineering*. Artech House.

Gustafsson, M., Jamal, K. and Dahlman, E. (September 1997). Compressed mode techniques for inter-frequency measurements in a wideband DS-CDMA system. *Proceedings of PIMRC '97*, Helsinki, Finland.

Heine, G. *GSM Networks*. Artech House.

Holma, H. and Toskola, A. (2000). *WCDMA for UMTS. Radio Access for 3rd Generation Mobile Communications*. John Wiley & Sons.

Jakes, W.C. (1974). *Microwave Mobile Communications*. John Wiley & Sons.

Lilja, H. (1996). Characterizing the effect of non-linear amplifier and pulse shaping on the adjacent channel interference with different data modulations. Licentiate Thesis. Oulu University, Finland.

Ojanpera, T. and Prasad, R. (1998). *WCDMA for 3rd Generation Mobile Communications*. Artech House.

Ojanpera, T. and Prasad, R. (2001). *WCDMA: Towards IP Mobility and Mobile Internet*, pp. 89, 101–102. Artech House.

Peterson, R.L.R., Ziemer, E. and Borth, D.E. (1995). *Introduction to Spread Spectrum Communications*. Englewood Cliffs, NJ: Prentice Hall.

Recommendation ITU-R M.687, (1998). Future public land mobile telecomms systems (FPLMTS).

Sato, S. and Amezaw, Y. (1997). Study on dynamic zone control for CDMA mobile radio communications. *Proceedings of ICC '97*, Vol. 2, Montreal, Canada.

Shapira, J. (November 1994). Microcell engineering in CDMA cellular networks. *IEEE Transactions on Vehicular Technology*, Vol. 43, No. 4.

Shapira, J. and Padovani, R. (1992). Spatial topology and dynamics in CDMA cellular networks. *Proceedings of VTC'92*, Denver, Colorado, USA.

Simon, M.K., Omura, J.K., Scholz, R.A. and Levitt, B.K. (1984). *Spread Spectrum Communications Handbook*. McGraw Hill, Inc.

Part III: Coverage and Network Planning

3GPP Technical Specification 25.101, UE Radio Transmission and Reception (FDD).

3GPP Technical Specification 25.104, UTRA (BS) Radio Transmission and Reception (FDD).

3GPP (1999(a)). Mandatory Speech Codec Speech Processing Functions, AMR Speech Codec: General Description (3G TS 26.101, Version 1.4.0).

3GPP (1999(b)). Mandatory Speech Codec Speech Processing Functions, AMR Speech Codec: General Description (3G TS 26.071, Version 3.0.1).

Anderlind, E. and Zander, J. (March 1997). A traffic model for non-real-time data users in a wireless network. *IEEE Communications Letters*, Vol. 1, No. 2.

ETSI/SMG-5 (1997). Selection Procedures for the Choice of Radio Transmission Technologies of the UMTS. UMTS 30.03, Version 3.0.0.

Faruque, S. (1996). *Cellular Mobile Systems Engineering*. Artech House.

Ganesh, R. and O'Byrne, V. (December 1997). Improving system capacity of a dual mode CDMA network. *International Conference on Personal Wireless Communications*, Mumbai, India.

Gray, S.D. and Kenney, T. (September 1997). A technique for detection AMPS intermodulation distortion in a IS-95 CDMA mobile. *Proceedings of PIMRC '97*, Vol. 2, Helsinki, Finland.

Hamalainen, S., Holma, H. and Toskala, A. (1996). Capacity evaluation of a cellular CDMA uplink with multiuser detection. *Proceedings of ISSSTA*.

Hamied, K. and Labedz, G. (April 1996). AMPS cell transmitter interference to CDMA mobile receiver. *Proceedings of VTC '96*, Vol. 3, Atlanta, Georgia.

Heikkinen, T. and Hottinen, A. (May 1998). Downlink power control and capacity with multi-antenna transmission. *Proceedings of VTC '98*, Ottawa, Canada.

Holma, H. and Toskala, A. (2000). *WCDMA for UMTS. Radio Access for Third Generation Mobile Communication*, p. 88. John Wiley & Sons.

Juntii, M. and Latva-aho, M. Multiuser receivers for CDMA systems in Rayleigh fading channels. *IEEE Transactions on Vehicular Technology*.

Lilja, H. (1996). Characterizing the effect of non-linear amplifier and pulse shaping on the adjacent channel interference with different data modulations. Licentiate Thesis. Oulu University, Finland.

Ojanpera, T., and Prasad, R. (2001). *WCDMA: Towards IP Mobility and Mobile Internet*, p. 130. Artech House.

Ojanpera. T, and Prasad, R. (2001). *WCDMA: Towards IP Mobility and Mobile Internet*, p. 173. Artech House.

Ojanpera, T., Prasad, R. and Harada, H. (May 1998). Qualitative comparison of some multiuser detector algorithms for WCMA. *Proceedings of VTC '98*, Ottawa, Canada.

Pizarroso, M. and Jimenez, J. (1995). Common Basis for Evaluation of ATDMA and CODIT System Concepts. MPLA/TDE/SIG5/DS/P/001/b1, MPLA SIG 5.

Recommendation ITU-R M.1225 (1997). *Guidelines for Evaluation of Radio Transmission Technologies for IMT-2000 Question*. ITU-R 39/8.

Sipila, K., Honkasalo, Z., Laiho-Steffens, J. and Wacker, A. (2000). A estimation of capacity and required transmission power of WCDMA downlink based on a downlink pole equation. *Proceedings of VTC 2000*, Tokyo, Japan.

Wacker, A.,Laiho-Steffens, J., Sipila, K. and Heiska, K. (1999). The impact of the base station sectorization on WCDMA radio network performance. *Proceedings of VTC '99*, Amsterdam, The Netherlands.

Wallace, M. and Walton, R. (1994). CDMA radio network planning. *Proceedings of IEEE VTC '94*, Stockholm, Sweden.

Westman, T. and Holma, H. (May 1997). CDMA systems for UMTS high bit rate services. *Proceedings of VTC '97*, Phoenix, Arizona.

Part IV: Quality and Network Planning

3GPP Technical Specifications TS 22.105 V.4.0.0 (2000). Services and Service Capability.

3GPP Technical Specifications TS 23.207 V.0.0.1 (2000). End-to-End QoS Concept and Architecture.

3GPP Technical Specifications TS 23.107 V.4.0.0 (2000). QoS Concept and Architecture.

Ariyavisitakul, S. (1992). SIR based power control in a CDMA system. *Globecom '92*, Orlando, Florida, USA.

Braden, R., Zhang, L., Berson, S., Herzog, S. and Jamin, S. (September 1997). Resource Reservation Protocol (RSVP). IETF RFC 2005.

Christer, B. and Johansson, V. (May 1998). Packet data capacity in wideband CDMA system. *Proceedings of VTC '98*, Ottowa, Canada.

Dahlman, E., Knutsson, J., Ovesjo, F., Persson, M. and Roobol, C. (November 1998). WCDMA – the radio interface for future mobile multimedia communications. *IEEE Transactions on Vehicular Technology*, Vol. 47, November 4.

ETSI UMTS 30.06 (December 1997). UMTS terrestrial radio access (UTRA): Concept Evaluation Version 3.0.0.

ETSI UMTS 30.03. (1997). Selection procedures for the choice of radio transmission technologies of the UMTS. *ETSI Technical Report*.

Handley, M., Schulzrine, H., Schooler, A. and Rosenberg, J. (April 2000). SIP: Session Initiation Protocol, draft-IETF-SIP-RFC2543bis-)1-pdf. IETF Draft, IETF.

Hanley, M. (1999). SIP: Session Initiation Protocol RFC2543 IETF.

Hashemi, H. (1993). Indoor propagation channel. *Proceedings of IEEE*, Vol. 18.

Holma, H. and Toskala, A. (2000). WDCMA for UMTS. *Radio Access for 3rd Generation Mobile Communications*, pp. 157–159, 166–170, 179, 185. John Wiley & Sons.

ITU-T H.323V2 (1998). Packet based multimedia communications systems.

ITU-T H.324 (1998). Terminal for low bit rate multimedia communications.

Jalali, A. and Mermelstein, P. (June 1994). Effects of diversity, power control and bandwidth on the capacity of micro cellular CDMA system. *IEEE Journal on Selected Areas in Common*, Vol. 12, No. 5.

Knutsson, J., Butovitsch, P., Persson, M. and Yates, R. (1998). Downlink admission control strategies for CDMA sytems in a Manhattan environment. *Proceedings of VTC '98*, Ottowa, Canada.

Molkdar, D. (February 1991). Review on radio propagation into and within buildings. *IEE Proceedings – H*, Vol. 138.

Proakis, J.G. (1995). *Digital Communications*, 3rd edition. McGraw-Hill.

Salonaho, O. and Laakso, J. (May 1999). Flexible power allocation for physical control channel in wideband CDMA. *Proceedings of VTC '99*, Spring, Houston, Texas, USA.

Rappaport, T.S. (1996). *Wireless Communication Principles and Practice*. Englewood Cliffs, NJ.

Recommendation ITU-R M.(FPLMTS.REVAL). *Guidelines for Evaluation of Radio Transmission Technologies for IMT-2000/FPLMTS.*

Tiedemann Jr, E.G. and Jou, J.C. and Odenwalder, J.P. (1997). The evolution of IS-95 to a third generation system and to the IMT-2000 era. *Proceedings of ACTS Summit*, Aalborg, Denmark.

Xia, H.H., Herrera, A.B., Kim, S. and Rico, F.S. (1996). A CDMA distributed antenna system for in-building personal communications services. *IEEE Journal on Selected Areas in Communications*, Vol. 14, No. 4.

Part V: Optimization and Network Planning

3GPP Technical Report 25.942. RF System Scenarios.

Anderson, J.B., Rappaport, T.S. and Yoshida, S. (1995). Propagation measurements for wireless communication channels. *IEEE Communications Magazine*, Vol. 33.

Cheung, J.C.S., Beach, M.A. and McGeehan, J.P. (1994). Network planning for 3rd generation mobile radio systems. *IEEE Communications magazine*, Vol. 32, No. 11.

Dahlberg, T.A., Ramaswany, S. and Tipper, D. (1997). Survivability issues in wireless mobile networks. *International Workshop on Mobile and Wireless Communications Networks.*

ETSI UMTS 30.03 (1997). Selection Procedures for the Choice of Radio Transmission Technologies of the UMTS. *ETSI Technical Report.*

Fleury, B.H. and Leuthold, P.E. (1996). Radiowave propagation in mobile communications: an overview of European research. *IEEE Communications Magazine*, Vol. 34.

Gibson, J.D. (1997). *The Communications Handbook.* CRC Press.

Hashemi, H. (1993). *Indoor Propagation Channel Proceedings IEEE*, Vol. 18.

Holma, H. and Toskala, A. (2000). WDCMA for UMTS. *Radio Access for 3rd Generation Mobile Communications*, pp. 166–167, 230. John Wiley & Sons.

Holma, H. and Toskala, A. (2000). WDCMA for UMTS. *Radio Access for 3rd Generation Mobile Communications,* pp. 194–195, 205–208. John Wiley & Sons.

Jakes, W.C. (1974). *Microwave Mobile Communications.* New York: Wiley & Sons.

Lee, J. and Miller, L. (1998). *CDMA Systems Engineering Handbook.* Artech House.

Miller, L.E. (July 1992). *Propagation Model Sensitivity Study.* J.S. Lee Associates, Inc, Report JC-2092-1-FF under contract DAAL02-89-C-0040, July.

Molkdar, D. (February 1991). Review on radio propagation into and within buildings'. *IEE Proceedings – H*, Vol. 138.

Ojanpera, T. and Prasad, R. (2001). *WCDMA: Towards IP Mobility and Mobile Internet*, pp. 250, 280–282, 272, 276–277. Artech House.

Ojanpera, T. and Prasad, R. (2001). *WCDMA: Towards IP Mobility and Mobile Internet*, p. 293. Artech House.

Ojanpera, T., and Prasad, R. (2001). *WCDMA: Towards IP Mobility and Mobile Internet*, pp. 347–348, 350, 366, 369. Artech House.

Rappaport, T.S. (1996). *Wireless Communication Principles and Practice.* Englewood Cliffs, NJ.

Recommendation ITU-R M.(FPLMTS.REVAL). *Guidelines for Evaluation of Radio Transmission Technologies for IMT-2000/FPLMTS.*

Sellakis, H. and Girdano, A. (1996). CDMA radio planning and network simulation. *IEEE International Symposium on Personal Indoor and Mobile Radio PIMRC '96*, Taiwan.

Singer, A. (February 1998). Improving system performance. *Wireless Review.*

Xia, H.H. (1997). Reference models for evaluation of third generation radio transmission technologies. *Proceedings of ACTS Summit*, Aalborg, Denmark.

Xia, H.H., Herrera, A.B., Kim, S. and Rico, F.S. (1996). A CDMA distributed antenna system for in-building personal communications services. *IEEE Journal on Selected Areas in Communications*, Vol. 14, No. 4.

Part VI: UMTS (3G) Development

ETSI website: http://www.etsi.org

Guntsch, A., Ibnkahla, M., Losquadro, G., Mazzella, M., Rovidas, D. and Timm, A. (February 1998). EU's R&D activities on 3rd generation mobile satellite systems (S-UMTS). *IEEE Personal Communications*, Vol. 36, No. 2.

Holma, H. and Toskala, A. (2000). *WCDMA for UMTS.* John Wiley & Sons.

IEEE 802.15 Working Group for Wireless Personal Area Networks. http:// grouper.ieee.org/groups/802/15.

ITU website: http://www.itu.int

LAN Medium Access control (MAC) and Physical Layer (PHS) pecs. (1997). IEEE Std. 802.11.

Ojanpera, T. and Prasad, R. (2001). *WCDMA: Towards IP Mobility and Mobile Internet*, pp. 444. Artech House.

Orthogonal Frequency Division Multiplexing. Magis Networks, Inc. www.magisnetworks.com

Prasad, R. (1996). *CDMA for Wireless Personal Communications*. Norwood, Massachusetts. Artech House.

Radio Equipment and Systems (RES). High Performance Radio Local Area Network (HIPERLAN) Type 1, Functional Spec. ETS 300 652.

Richard, D.J. and Prasad, R. (2000). *OFDM for Wireless Multimedia Communications*. Artech House.

UMTS Forum website: http://www.umts-forum.org

UMTS (The) Markets Aspect Group (1997) *UMTS Market Forecast*, Dtufy.

Westman, T. and Holma, H. (1997). CDMA system for UMTS high bit rate services. *Proceedings of VTC '97*, Phoenix, Arizona.

General Bibliography

Kim, K. (2000). *Handbook of CDMA System Design, Engineering and Optimization*. Prentice Hall.

Laiho, J., Wacker, A. and Novosad, T. (2002). *Radio Network Planning and Optimization for UMTS*. John Wiley & Sons.

Lee and Miller (1998). *CDMA Systems Engineering*. Artech House.

Maylor, H. (2003). *Project Management*. Prentice Hall.

Muhlemann, A., Oakland, J. and Lockyer, K. (1992). *Production and Operation Management*. Pitman.

Oakland, J.S. (1993). *Total Quality Management*, 2nd edition. Butterworth Heinemann.

White, G. (1994). *Mobile Radio Technology*. Butterworth Heinemann.

Yang, S.C. (1998). *CDMA RF System Engineering*. Artech House.

Glossary: UMTS Acronyms

This part contains a list of third-generation (3G)-related acronyms presented in this document. For more detailed acronym lists, please refer to the 3G partnership projects (3GPP) specifications:

1G	the first (cellular) generation
2G	the second (cellular) generation
2.5G	2.5G (GPRS)
3G	the third (cellular) generation
3GPP	third-generation partnership project
3GPP1	third-generation partnership project 1
3GPP2	third-generation partnership project 2
3GPP MAP	enhanced MAP version for 3GPP purposes
4G	the fourth (cellular) generation

A

A	BSC–MSC interface
AAL(n)	ATM adaptation layer n, where $n = 1 \ldots 5$
Abis	BSC–BTS interface
ABR	available bit rate service
AC	authentication centre
ACI	adjacent channel interference
ACIR	adjacent channel interference ratio
ACL	adjacent channel leakage
ACLR	adjacent channel leakage ratio
ACM	address complete message
ACTS	advanced communications technologies and services
AGC	automatic gain control
AGL	above-ground level
AH	address handling
AICH	acquisition indication channel
ALERT	alerting message
AMPS	American Mobile Phone System
AMR	adaptive multi-rate
ANM	answer message

ANSI	American National Standard Institute
ARIB	Association of Radio Industries and Business
ARPU	average revenue per user
ATM	asynchronous transfer mode
AuC	authentication centre
AUTN	authentication token
AVN	authentication vector array

B

BCCH	broadcast control channel
BCH	broadcast channel
BER	bit error ratio
Bi	information bandwidth
BIB	backward indicator bit
BLO	block message
BS	base station
BSC	base station controller (GSM)
BSN	backward sequence number
BSS	base station subsystem
BSSAP	base station subsystem application part
Bt	transmission bandwidth
BTS	base transceiver station

C

CAMEL	customized applications for mobile network enhanced logic
CBR	constant bit rate service
CC	call control/country code
CCCH	common control channel
CCF	call control function
CCPCH-1	primary common control physical channel
CCPCH-2	secondary common control physical channel
CCQ	coverage, capacity, and quality (of service)
CCS7	common channel signalling number 7
CCITT	Consultative Committee for International Telegraph and Telephone
CDG	CDMA development group
CDMA	code division multiple access

cdma2000 code division multiple access 2000
CDR call data record
CE channel element
CFN confusion message
CGI cell global identity
CHG charge message
CI cell identity
CIC circuit identification code
CS check sum
CK cipher key
CLP cell loss priority
CM communication management
CN core network
CN-CS core network circuit-switched domain
CN-PS core network packet-switched domain
CODIT code division test bed
COSSAP communications simulation and system analysis program
CPCH common packet channel
CPICH common pilot channel
CPICH-1 primary common pilot channel
CPICH-2 secondary common pilot channel
CRC cyclic redundancy check
CS convergence sublayer
CSCF call state control function
CSE CAMEL service environment

D

DAB digital audio broadcast
dB decibel
dBi decibel (reference to an isotropic radiator)
DCA dynamic channel allocation
DCCH dedicated control channel
DCH dedicated channel
DDD data delivery density
DECT digitally enhanced cordless telephone
DNS domain name server/domain name system
DPC destination point code

DPCCH	dedicated physical control channel
DPCH	dedicated physical channel
DPDCH	dedicated physical data channel
DRNC	drift RNC
DS	direct sequencing
DSP	digital signal processor
DTCH	dedicated traffic channel
DVB	digital video broadcasting
DVB-T	digital video broadcasting – terrestrial

E

Eb	energy per bit
Ec	energy per chip
EDGE	enhanced data rates for GSM evolution/environment
EIR	Equipment Identity Register
EIRP	effective isotropic radiated power
Erl	Erlangs
ERP	effective radiated power
ETSI	European Telecommunication Standard Institute

F

F	frame mark (CCS7)
FACH	forward access channel
FDD	frequency division duplex
FDMA	frequency division multiple access
FER	frame error rate
FFT	fast Fourier transform
FH	frequency hopping
FIB	forward indicator bit
FISU	fill-in signalling unit
FPGA	field programmable gate arrays
FSN	forward sequence number

G

GDO	general development order
GFC	generic flow control
GGSN	gateway GPRS support node
GHz	gigahertz

GMSC	gateway MSC (could be functionality or equipment)
GoS	grade of service
Gp	processing gain
GPRS	general packet radio service
GPS	global positioning system
GSM (old)	groupé specialé mobile
GSM	global system for mobile communications
GTP	GPRS tunnelling protocol

H

h	hour
HCS	hierarchical cell structure
HDR	high data rate
HEC	header error control
HiperLAN	hiper local area network
HiperLAN2	hiper local area network with extended inter-connections
HLR	Home Location Register
HSCSD	high-speed circuit-switched data
HSS	home subscriber server
html	hypertext markup language

I

IAM	initial address message
ICGW	incoming call gateway
IDFT	inverse discrete Fourier transform
IEEE	Institute of Electrical and Electronic Engineers
IETF	Internet Engineering Task Force
IF	intermediate frequency
IK	integrity key
Im	interference margin
IM	intermodulation
IMAP	Internet message access protocol
IMEI	international mobile equipment identity
IMS	IP multimedia sub-system
IMSI	international mobile subscriber identity
IMT-2000	international mobile telephony 2000
IMUI	international mobile user identity
IN	intelligent network

INAP	IN application part
IP	Internet protocol
IPSec	IP security
IPv4	Internet protocol Version 4
IPv6	Internet protocol Version 6
IS-95	cdmaOne system
IS-136	TDMA system (Second generation)
ISDN	integrated services digital network
ISI	inter-symbol interference
ISM	industrial scientific and medical bands
ISP	Internet service provider
ISS	instruction set simulators
ISUP	ISDN user part
IT	information technology
ITU	International Telecoms Union
ITU-T	International Telecommunication Union – Telecommunication Standardization Sector
Iu	Interface between RAN and CN
Iub	Interface between RNC and Node-B
Iur	Interface between RNC and RNC

K

kbps	kilobits per second

L

LA	location area
LAC	location area code
LAI	location area identity
LAN	local area network
LDMOS	laterally diffused metal oxide semi-conductor
LI	length indicator
LIF	Location Interoperability Forum
LNA	low-noise amplifier
LOS	line of site
LSSU	line status signalling unit
LU	location update

M

MAC	medium access control
MAP	mobile application part (protocol)
Mbph	megabits per hour
Mbps	megabits per second
MBS	mobile broadband system
MC	multi-carrier
MCC	mobile country code
MC-CDMA	Multi-carrier CDMA
MCM	Multi-carrier modulation
Mcps	megachips per second
MEMS	micro-electrical–mechanical systems
mErl	milli Erlangs
MexE	mobile executing environment
MGCF	media gateway control function
MGW	media gateway
MM	mobility management
MMM	mobile multi-media messaging
MMSE	minimum mean square error
MNC	mobile network code
MPLS	multi-protocol label switching
MPM	metering pulse – message
ms	millisecond
MS	mobile station
MSC	mobile-switching centre
MSISDN	mobile subscriber ISDN number (directory number)
MSN	mobile subscriber number
MSRN	mobile subscriber roaming number
MSU	message signalling unit
MT	mobile terminal
MTP	message transfer part
MUD	multi-user detection

N

NBAP	node B application protocol
NDC	national destination code
NLOS	no line of site

NMS	network management system, see OSS and OMC
NMT	Nordic Mobile Telephone
No	noise spectral density
Node B	base station (3G)
NRT	non-real time
NSS	network subsystem (circuit switching for GSM)

O

OFDM	orthogonal frequency division multiplexing
OHG	Operator Harmonization Group
OMC	operation and management centre
OPC	originating point code
OQPSK	offset quadrature phase shift keying – RF modulation method
OSA	open system architecture
OSI	open system interconnection
OSS	operating subsystem, see NMS and OMC
OTOR	omni-transmit and omni-receive
OTSR	omni-transmit and sector receive
OVSF	orthogonal variable spreading factor

P

PCH	paging channel
PCM	pulse code modulation
PCPCH	physical common packet channel
PDA	personal digital assistant
PDC	personal digital communication system
PDH	plesiosynchronous hierarchy
PDP	packet data protocol
PDSCH	physical shared data channel
PICH	paging indicator channel
PRACH	physical random access channel
P-SCH	primary synchronization channel
PSTN	public-switched telephone network
PT	payload type
P-TMSI	packet TMSI

Q

QoS	quality of service
QPSK	quadrature phase shift keying – RF modulation method

R

RA	routing area
RAB	radio access bearer
RACH	random access channel
RAKE	type of receiver utilizing multi-path propagation
RAN	radio access network
RANAP	radio access network application protocol
RAND	random number
RAU	routing area update
RCM	release complete message
REL	release message
RES	resume message
RF	radio frequency
RLC	radio link control
RNC	radio network controller
RNS	radio network subsystem
RNSAP	radio network subsystem application protocol
RRC	radio resource control
RRM	radio resource management
RSC	reset circuit message
RSVP	reservation protocol
RT	real time
RTCP	real-time control protocol
RTP	real-time protocol
RTT	radio transmission technology
RX	receiving functionality/receiver

S

s	seconds
SABM	set asynchronous balanced mode
SAMBA	system for advanced mobile broadband applications
SAPI	service access point identifier
SAR	segmentation and reassembly sublayer

SAT	SIM application toolkit
SCCP	signalling connection control part
SCH	synchronization channel
SCH-1	primary synchronization channel (physical)
SCH-2	secondary synchronization channel (physical)
SCP	service control point (for intelligent network)
SDH	synchronic digital hierarchy
SDMA	space division multiple access
SF	spreading factor
SFN	system frame number
SGSN	serving GPRS support node
SIF	service information field
SIM	subscriber identity module
SIO	service information octet
SIP	session initiation protocol
SIR	signal-to-interference ratio
SLS	signalling link selection
SM	session management
SMS	short message service
S-MSC	serving MSC
SMSC	short message service centre
SN	subscriber number
S/N	signal to noise
SNMP	simple network management protocol
SNR	signal-to-noise ratio
SPC	signalling point code
SPD	serving profile database
SRES	signed response
SRNC	serving RNC
SS	supplementary services
SSP	service-switching point (for intelligent network)
STM	synchronous transport module
STM-1	synchronous transport module number 1
STM-4	synchronous transport module number 4
STM-16	synchronous transport module number 16
STM-64	synchronous transport module number 64
STP	signalling transfer point
SUS	suspend message

T

TCAP	transaction capabilities application part
TCP	transmission control protocol
TDD	time division duplex
TDMA	time division multiple access
TE	terminal equipment
TFCI	transport format combination indicator
TFTP	trivial file transfer protocol
TMA	tower-mounted amplifier
TMSI	temporary mobile subscriber identity
TMUI	temporary mobile user identity
TOS	type of service
TRAU	transcoder and adaptation unit
TRX	transmitter and receiver
TTA	Telecoms Technology Association
TTP	traffic termination point
TX	transmitting functionality/transmitter

U

UA	unnumbered acknowledgement
UBL	unblock message
UBR	unspecified bit rate service
UCIC	unknown CIC message
UDP	unit data protocol
UE	user equipment
Um	BSS–MS interface
UMTS	universal mobile telephony system
URL	unified resource locator
USIM	UMTS SIM
UTRA	universal terrestrial radio access
UTRAN	universal terrestrial radio access network
Uu	Interface between the UE and Node-B

V

VAS	value-added services
VBR	variable bit rate service

VCI	virtual channel identifier
VHE	virtual home environment
VLR	Visitor Location Register
VMS	voice mail system
VoIP	voice over Internet protocol
VPI	virtual path identifier
VPN	virtual personal network
VSWR	variable standing wave ratio

W

W-CDMA	wideband code division multiple access
WEP	wired equivalent privacy
WLAN	wireless local area network
WLL	wireless local loop

X

XRES	expected response

Index

D